CAMBRIDGE LIBRARY COLLECTION

Books of enduring scholarly value

Earth Sciences

In the nineteenth century, geology emerged as a distinct academic discipline. It pointed the way towards the theory of evolution, as scientists including Gideon Mantell, Adam Sedgwick, Charles Lyell and Roderick Murchison began to use the evidence of minerals, rock formations and fossils to demonstrate that the earth was older by millions of years than the conventional, Bible-based wisdom had supposed. They argued convincingly that the climate, flora and fauna of the distant past could be deduced from geological evidence. Volcanic activity, the formation of mountains, and the action of glaciers and rivers, tides and ocean currents also became better understood. This series includes landmark publications by pioneers of the modern earth sciences, who advanced the scientific understanding of our planet and the processes by which it is constantly re-shaped.

Seismology

While living in Japan, John Milne (1850–1913) sought to study the 1880 Yokohama earthquake, soon realising that scientists lacked the proper tools. Aided by colleagues, he went on to develop the necessary instrumentation, and by 1896 he had built the first seismograph capable of recording major earthquakes in any part of the world. His textbook *Earthquakes and Other Earth Movements* (also reissued in this series) had appeared in 1886. In this follow-up work, published in 1898, Milne continues to discuss the nature of earthquakes, the methods and equipment needed to investigate them, and how to apply this knowledge to construction. He references the research, hypotheses and formulae of modern scientists, also noting in passing the suggestions made by earlier authors on the causes of seismic activity. The text is accompanied by many diagrams, especially of experimental apparatus, and several photographs illustrate damaged buildings and bridges.

Cambridge University Press has long been a pioneer in the reissuing of out-of-print titles from its own backlist, producing digital reprints of books that are still sought after by scholars and students but could not be reprinted economically using traditional technology. The Cambridge Library Collection extends this activity to a wider range of books which are still of importance to researchers and professionals, either for the source material they contain, or as landmarks in the history of their academic discipline.

Drawing from the world-renowned collections in the Cambridge University Library and other partner libraries, and guided by the advice of experts in each subject area, Cambridge University Press is using state-of-the-art scanning machines in its own Printing House to capture the content of each book selected for inclusion. The files are processed to give a consistently clear, crisp image, and the books finished to the high quality standard for which the Press is recognised around the world. The latest print-on-demand technology ensures that the books will remain available indefinitely, and that orders for single or multiple copies can quickly be supplied.

The Cambridge Library Collection brings back to life books of enduring scholarly value (including out-of-copyright works originally issued by other publishers) across a wide range of disciplines in the humanities and social sciences and in science and technology.

Seismology

JOHN MILNE

CAMBRIDGE
UNIVERSITY PRESS

CAMBRIDGE
UNIVERSITY PRESS

University Printing House, Cambridge, CB2 8BS, United Kingdom

Cambridge University Press is part of the University of Cambridge.

It furthers the University's mission by disseminating knowledge in the pursuit of education, learning and research at the highest international levels of excellence.

www.cambridge.org
Information on this title: www.cambridge.org/9781108070294

© in this compilation Cambridge University Press 2014

This edition first published 1898
This digitally printed version 2014

ISBN 978-1-108-07029-4 Paperback

This book reproduces the text of the original edition. The content and language reflect the beliefs, practices and terminology of their time, and have not been updated.

Cambridge University Press wishes to make clear that the book, unless originally published by Cambridge, is not being republished by, in association or collaboration with, or with the endorsement or approval of, the original publisher or its successors in title.

THE

INTERNATIONAL SCIENTIFIC SERIES

VOL. LXXXV.

SEISMOLOGY

BY

JOHN MILNE, F.R.S., F.G.S.

LATE PROFESSOR OF MINING, GEOLOGY, AND SEISMOLOGY IN THE
IMPERIAL COLLEGE OF ENGINEERING, TOKIO, JAPAN
HON. FELLOW KING'S COLLEGE, LONDON

WITH FIFTY-THREE FIGURES

FIRST EDITION

LONDON
KEGAN PAUL, TRENCH, TRÜBNER & CO. LTD.
PATERNOSTER HOUSE CHARING CROSS ROAD
1898

INTRODUCTION

ALTHOUGH several chapters in the present volume bear
the same title as those in a work upon Earthquakes
and other Earth Movements written for the 'International
Scientific Series' in 1883, it will be found that the subject
matter of these largely consists of observations which are
not only new, but more extensive and trustworthy than
were formerly obtainable.

The result of this is that the conclusions which are
formulated are not only more definite than those hitherto
arrived at, but are in many instances novel in character.

The chapters dealing with changes in the vertical,
diurnal waves, earth pulsations, the unfelt earthquakes so
common in all countries, indicate that movements of the
earth's crust can be equally well recorded and studied in
England and other non-volcanic countries as in the most
frequently earthquake-shaken districts in the world.

The records of these ubiquitous breathings of the
earth's surface, the observation of which is at present confined to one or two observers, constitute a new departure

in an old study, and promise to throw new light upon the physics of our earth's crust and the nature of its interior.

The practical outcome of seismometry will be found in chapters relating to construction in earthquake countries, which, by kind permission, are largely reproduced from articles contributed by me to ' Engineering,' and in a special chapter on the relationship of earth movements to the work of astronomers, physicists, assayers, colliery viewers, and those who construct and work deep sea cables. A few remarks have been added upon the recording of artificially produced vibrations and movements which so often have a marked existence in trains, locomotives, bridges, and steamships.

Although the references to work carried out in Italy and Germany are numerous, a recent visit to observatories in those countries has shown me that these might have been extended. Publications in which this work is described are however given in the text.

For assistance in the compilation of this volume my thanks are due to many. The Royal Geographical Society, the British Association and ' Engineering ' have allowed me to reproduce many of their blocks. To Mr. W. K. Burton and Dr. F. Ōmori I am indebted for several original photographs. Mr. C. D. West has often helped me in the designing of instruments and in actual observation, whilst other colleagues in Japan, the authorities of the Imperial University and the Meteorological Bureau in that country have always furnished me with a helping hand. For assistance in the revision of proofs, and especially for additions relating to earthquake periodicity, I am greatly indebted to Dr. C. G. Knott. In short, when I look back and think of the kindly advice I have during the last twenty years received from Lord Kelvin, Professor John Perry, and other members of committees for which I have worked, when I remember the liberality of the Royal Society, the British Association and the Geological Society, and when I recall the work of J. A. Ewing, Thomas Gray, G. H. Darwin, C. Davison, Sekiya, Yamakawa, Fujioka, von Rebeur-Paschwitz, and the many others through whose labours I have benefited, I feel that the results brought together in this little volume illustrate the value of co-operation among scientific workers.

JOHN MILNE.

SHIDE, NEWPORT, ISLE OF WIGHT.
June 1898.

CONTENTS

—◦◦—

CHAPTER I

BRADYSEISMS

CHAPTER II

METHODS OF MEASURING BRADYSEISMICAL MOTION

CHAPTER III

CAUSES OF EARTHQUAKES

CHAPTER IV

SEISMOMETRY

CHAPTER V

THE NATURE OF EARTHQUAKE MOTION

CHAPTER VI

VELOCITY OF EARTH WAVES

CHAPTER VII

SEISMIC ELEMENTS WHICH ARE CALCULABLE

CHAPTER VIII

EARTHQUAKES AND CONSTRUCTION

CHAPTER IX

EARTHQUAKES AND CONSTRUCTION (*continued*)

CHAPTER X

THE POSITION, CHARACTER, DEPTH, AND DISTRIBUTION
OF EARTHQUAKE ORIGINS

CHAPTER XI

SEISMIC FREQUENCY AND PERIODICITY

CHAPTER XII

SEISMIC PHENOMENA OF A MISCELLANEOUS CHARACTER

CHAPTER XIII

SLOW CHANGES IN THE VERTICAL

CHAPTER XIV

THE DIURNAL AND SEMI-DIURNAL WAVES

CHAPTER XV

PULSATIONS

CHAPTER XVI

EARTH TREMORS

CHAPTER XVII

MOVEMENTS OF THE EARTH'S CRUST IN RELATION TO PHYSICAL RESEARCH AND ENGINEERING

SEISMOLOGY

———•◇•———

CHAPTER I

BRADYSEISMS

Insignificance of irregularities on the Earth's surface relatively to its
size—Bradyseismical action in Japan—Variations in the height of
mountains—The exact height of a mountain is not determinable—
The want of fixity in the datum relatively to which Bradyseisms are
measured—Movements of water level in a basin by movements of its
boundaries—Effects of change of slope on ocean coasts upon the
advance or retreat of water—Change in water level due to the emergence of continental areas—generally, and at different geological
epochs—A large percentage of what is usually considered due to
rising of the land may be due to the falling of the water—Buckling
of strata on a seaboard may be accompanied by conditions such as are
evidenced by the coal measures—The uplifting of great mountains has
therefore accompanied the formation of coal—At these times volcanic
and seismic activity should have been marked.

In comparison with ourselves our world is large, its
mountains and valleys are gigantic excrescences on its
surface, whilst the elevations and depressions, representing
continental elevations and ocean basins, form irregularities
the magnitude of which we can only appreciate by the aid
of figures.

Directly, however, we compare these deviations from
smoothness with the world itself, we are astonished at their
insignificance.

On a model of our globe one hundred feet in diameter,
mountains and oceans which take travellers many days to
pass only appear as small ridges and gentle depressions,
and we are disappointed by their smallness.

B

If the diameter of the model is reduced to a foot, features which form the grandest scenery or basins forming the largest oceans may be represented by the almost imperceptible puckerings and depressions produced on a film of varnish which had dried upon its surface. Ocean depressions and continental elevations would be practically invisible, and we might pass our hand round and round the model without noticing any irregularity. It is doubtful whether any molten sphere of metal like such a model would, after cooling, show less deviation from smoothness than those observed upon the surface of our earth.

If we therefore accept the idea that the excrescences upon the surface of our earth are in relation to its magnitude extremely slight, and add to this the idea that rocks in extended masses are capable of being bent and folded, rather than find difficulty in imagining that the surface irregularities of our sphere are due to a layer of rocks which is unable to support its own weight, accommodating itself to a contracting nucleus, we have much greater difficulty in realising why these irregularities have not been greater than we find them.

Before entering into a discussion of the relative importance of radial and lateral contraction, the effects of weight due to accumulating sediment, heating and cooling as evidenced by the rise or fall of isothermal surfaces, the compression accompanying the intrusion of volcanic dykes, chemical changes producing alteration in volume, and other influences which may have played a part in the production of terrestrial features, we make it our first object to give a few new illustrations taken from the coast of Japan which indicate that bradyseismical changes are yet in operation.

In 1891 Professor D. Kikuchi, of the Imperial University of Japan, issued a circular to officials at the principal towns and villages round the coast of that country requesting them to forward any evidence that they were able to collect which showed that there had been encroachment or recession of water on the seaboard. Several thousand replies were received, and it was but very few of these which indicated that there had been no change.

At many places during the last fifty years, and in some cases even ten years, we learned that harbours had grown shallower at rates varying from one foot in three years to one foot in ten years. Only small vessels are now able to enter these harbours, rocks which were beneath the surface are now above the water, posts to which ships were fastened are now 180 feet inland; shallow wells pass through beds of shells like those at present in the sea; the mouths of rivers have grown shallower; the tide leaves a greater area of coast bare than it did in former years; fishermen who had placed their nets at a distance of 1,200 feet from the shore, have now to go a distance of 1,800 feet to find water of a similar depth. These and other facts point to the conclusion that at many places—as, for example, round the Shimabara Gulf, in the Inland Sea, on the coast of Sagami, to the north and south of Sendai, and generally on the eastern and southern sides of Japan—elevation has been taking place within the memory of the living.

From other places we learn that grass and rice fields are now represented by beaches of sand or shingle; that the depth of the sea has increased at rates of from one foot in sixteen years to one foot in five years; that rocks have sunk, and the height of the tide has increased; that buildings are nearer to the water line than they were when first erected, the water in some cases approaching roads and buildings so rapidly that the inhabitants are contemplating moving inland; that maps of a hundred years ago show sites of former dwelling-places that are now beneath the sea. Although in Japan submerged forests or 'dirt beds' are unknown, we find, on the west coast near Iwanai and on the shores of Kaga, submarine depressions following the line of valleys on the land. All these facts point to the conclusion that certain districts, especially those to

the north of Noto bordering the China Sea, are slowly
sinking.

Those who describe these changes usually attribute
them to the accumulation of sediment, the washing away
of coast material, or to the occurrence of some great
earthquake, although in no case has it been stated that the
changes accompanied such disturbances.

By taking a series of maps representing the Tokyo
district, the first of which dates from the year A.D. 1028,
and superimposing them one upon the other, we can
readily determine the average rates at which the ground
on which the present city is built has grown seawards.
At one point the average rate has been thirty-eight feet per
year, while at another it has been only two feet per year.

During a residence of nearly twenty years in Tokyo, I
have seen mud banks appear which now have been re-
claimed, so that the area of the ground bordering the sea
frontage has been increased by *very many* acres.

These changes are no doubt largely due to deposition
of sediment brought down by the Sumida and other rivers
entering the bay ; but when we look at the shell borings
in the rocks flanking this sheet of water. we are compelled
to admit that this rapid shallowing and growth of land
must at least in part be due to actual elevation.   As one
example of these shell borings, I may mention several lines
of them in the cliff forming the face of the Bluff at
Yokohama. The rock is a soft clayey tuff, and the borings
are to be seen in this at a height of about ten feet above
high-water mark.   Because this rock is so extremely soft
and easily acted upon by the weather. it is difficult to
suppose that the borings can have been formed more than
fifty years ago. If, however, we double this limit and
exclude paroxsysmal action, we are led to the belief that
elevation has been going on at a rate of one foot in ten
years, a rate quite comparable with those obtained along
coast lines which have been already mentioned. Eighteen
years ago, near the site of these markings, a point of rock
projected into the sea which the author does not remember

ever having been then able to pass. For the last few years
at low water he has passed it repeatedly, walking on what
is practically a rocky surface. At the lowest estimate
these observations would indicate that at many places on
the coast of Japan land has been emerging from the
waters at the rate of about one inch per year.

Round the shores of Japan, especially upon the south-
western coast of Yezo, sea-worn caves and hollows, raised
beaches and terraces, are evidence of more extended elevation. Near Hakodate and from Matsumai towards the
north, these latter, which are from twenty to forty feet in
height, are so well defined that they attract the attention
of passengers on passing steamers. On the western side of
Iterup the first terrace, which is half a mile or so in breadth,
has a face about 130 feet in height. From 200 to 300 feet
or so above this, the level of a second terrace is reached.

Here, as on the eastern side of the Pacific, it will be
observed that as we travel northwards traces of ancient
shore-lines occur at higher levels, and what is generally
true for northern latitudes is generally true as we proceed
down the coast of Peru and Chili for a distance of more
than 2,000 miles towards Valparaiso. At the latter place
Darwin found such indications at an elevation of 1,800
feet, while A. Agassiz found corals attached to rocks at a
height of 3,000 feet.

Although the above illustrations have been drawn
from Japan and Pacific coasts, similar illustrations of the
instability of the land relative to the surface of a neighbouring sea or ocean can be seen along nearly all the sea
coasts of the world, the most striking, perhaps, being
those which have taken place in the historical period, and
even within the memory of man. In some instances the
movement has not been altogether in one direction—
which is perhaps one of the most remarkable features
connected with these phenomena—but, as in the well
known case of the temple of Jupiter Serapis since the
Roman Period, an area has been depressed some twenty
feet and then elevated to its old position.

From these few notes it will be gathered that the rate
at which bradyseismical change takes place is extremely
variable. Sometimes evidence of the same may be inferred
from changes which are said to have taken place in the
relative heights of hills and mountains.

Sir Richard Worsley, in his ' History of the Isle of
Wight,' written in 1781, tells us that Shanklin Down
now stands about 100 feet higher than Week Down, yet
old persons affirm that Shanklin Down was formerly
hardly visible from St. Catherine's, also they knew when
Shanklin Down could not be seen from Chale Down, but
only from the top of the beacon. From this it would
appear that the intermediate Down (Week Down) has
sunk, or one of the other hills has risen.

In previous publications the writer has given several
illustrations of observations which closely resemble that
which is supposed to have taken place at Week Down
(' Earthquakes,' p. 352, International Scientific Series).
Unless these movements are pronounced, the difficulties
which surround the accurate determination of the height of
any fairly high mountain will render the measurement of
such changes almost an impossibility. The late T. W.
Blakiston, R.A., showed that fourteen observers who
endeavoured to measure the height of Mount Fuji gave
results varying between 11,000 and 14,000 feet. The
methods followed were based upon barometrical, thermometrical and hygrometrical observations extending over
one or two weeks made on the top of the mountain, and
simultaneously at two or three stations round its base,
hypsometrical determinations, a height measured by actual
levelling, and the results obtained by trigonometrical
observations. A point not to be lost sight of in connection with barometrical and hygrometrical observations is
that they lead to results which differ with the formula
employed in computation. For the particular mountain
considered, the conclusion arrived at was that in August
1884 the height of Mount Fuji was between 12,400 and
12,450 feet.

Because fourteen years have elapsed since these determinations were made, and because there are various causes known to geologists and to those who have studied the movements of the earth's crust which may possibly lead to changes in the height of a mountain, a careful re-measurement of Fuji would carry with it great interest ('Trans. Seis. Soc.' vol. xiv. pt. ii. p. 72).

I learn from Col. J. Farquharson, R.E., Director of the Ordnance Survey, that some years ago the question whether during recent years there had been any changes in level in Britain was carefully tested in Lancashire and Yorkshire, under the direction of Sir Charles Wilson. The first levelling in these counties was carried out between 1843 and 1850, and the second between 1888 and 1894. Excepting in the coal and salt districts, no material changes were found to have taken place. It is, however, to be remembered that this re-levelling was confined to lines of level along roads, and whether there have or have not been any changes in the height of hills or mountains since the first measurements were made we do not at present know.

For illustrations of a sudden seismical effect by which valleys have been compressed and mountains altered in height, the reader is referred to the various notes describing the great Japanese earthquake of 1891.

Again in 1897 a line of levels was run from Blackgang to Freshwater in the Isle of Wight, through St. Catherine's Tower, which stands on a chalk hill 781 feet in height. The level of the beach mark on the Tower agreed exactly with that obtained in 1853.

Appearances like raised sea-beaches and terraces indicate either that there have been times when the land moved more rapidly than usual, or that there have been comparatively rapid falls in sea level. The faulted and crumpled sedimentary strata which can be traced from the coast inland point to the fact that whatever movements there may have been in sea level, there have certainly been enormous movements in the rock.

The datum relatively to which elevations or depressions of the surface of our earth are usually measured is that of sea level. It must not, however, be overlooked that sea level, distorted, as it is, by gravitational and other effects, is in every probability a surface that has suffered many changes. By the accumulation of detritus deposited from rivers or thrown out from volcanic vents, and by the escape and condensation of vapours from submarine volcanic rocks, sea level may have been gradually raised.

The absorption of water by rocks would cause the same to fall. A diminution in the earth's rate of rotation would cause a fall of water in equatorial and a rise in polar regions.

These are actions which during long periods of time may have had at any one point on the earth's surface a resultant and fairly steady effect in one direction. Superimposed on this there may have been, at intervals corresponding to the recurrence of glacial epochs, risings and fallings, accompanying changes in the position of the earth's centre of attraction, and the distortional effects of the glacial load.

These steady and gradual changes in oceanic level are, however, inadequate to explain the *oscillatory* changes of sea or land, of which geologists have abundant evidence and which are often of a local character.

Changes in wind and barometrical pressure which at the most continue over a few days produce slight fluctuations in the level of a sea.

We are told that the difference between the summer and winter distances to which the Black Stream, a current comparable with the Gulf Stream, is felt as it runs north-eastwards along the coast of Japan is about 500 miles.

Inasmuch as the existence of ocean currents indicates that the oceans are bodies of water seeking a position of equilibrium, fluctuations in their velocity imply relative changes in oceanic level. If, therefore, we can assure ourselves of periodical changes in climatic conditions, we may infer corresponding changes in oceanic circulation and water level.

In a river channel with a given breadth and depth, a given velocity indicates a calculable difference in head between given points, but with an ocean current the elements required for a similar calculation have uncertain values.

One concomitant of variability in the velocity of ocean currents might possibly be detected at the entrances of certain gulfs.

In cases where such entrances were narrow, any increase in the velocity of the current might mean a rise in the waters in the bay, but a difference in level produced in this manner, even by a current which reached eight feet per second, could not exceed a foot.

We will now turn our attention to movements which may take place in the level of bodies of water accompanying movements of the land, the first illustration being that of a lake or inland sea. If the lateral ridges confining such a body of water are steep, so that the cross-section of the basin is V-shaped, the sinus of the depression containing the water forming with its boundaries a right angle, or an angle less than a right angle, then lateral compression, causing an actual elevation of the boundary ridges might, because of the accompanying *relative rise* in the confined waters, appear to produce a *depression* of these ridges. On the other hand, let the boundaries meet at an angle greater than a right angle and the water lie in a dish-like hollow. In this case lateral compression would result in a large *relative fall* in the water level, and the land elevation, as measured from the water datum, would be greatly exaggerated.

These cases of valley compression are by no means altogether hypothetical. Geologists have innumerable illustrations of valleys running in synclinal troughs, whilst the sudden compression of the Neo valley (Japan) in 1891 showed the direction rock-folding was following.

Again, we have no means of determining whether the bottoms of the lake-covered valleys have moved upwards, downwards, or have remained stationary, so that it is

impossible to give absolute measurements of vertical displacements. The vertical distance between the bottom of
a valley and its boundary ridges, and the horizontal
distance between two points on opposite ridges, may have
slightly changed, whilst large changes may have taken
place between either of these and the level of the sheet
of water they contain.

From a sheet of water that is closed we will turn to
an open ocean, the cross-section of which is that of an
extremely shallow dish. A slight change in the average
slopes of such a depression, although the effect would be
distributed over all other oceans, would result in the
advance or retreat of water from considerable areas on all
gently sloping shores. With basins of given form such
movements are calculable, and the magnitude may be
realised by the following simple experiment. Take a
board, say ten feet long, with a groove along its length.
Let this be supported at its two extremities, and the groove
be filled with water. The weight of the board together
with that of the water it carries will cause it to slightly
sag. By placing a small weight on the board the sag may
be increased, say one millimetre, and it will then be
observed that the water will run towards the centre for a
distance of thirty or forty millimetres. In this experiment the sag in the board is an exaggerated representation of a depression in the earth's crust, like an
ocean basin, and since the horizontal movement of the
water in the model is at least thirty times that of the
vertical displacement, it is not difficult to appreciate how
great the ratio might be in any real case of change of
slope.

The only conclusion to which these considerations lead
us is that water may rise or fall with changes in the form
of its containing basin, and so cover or expose large areas
of land.

The next object is to extend the inquiry respecting
possible fluctuations in the position of water level to a
consideration of how far such a datum may have changed

by the growth of continental areas and at certain epochs in geological history.

In the first case, one assumption is that in the history of Geography there was a period when the globe, whatever its configuration may have been, was nearly, if not completely, surrounded by water.

If the idea of extended tumefaction in the crust of such a globe is excluded as a physical impossibility, any deformation in the crust unaccompanied by protrusion above the surface of the liquid envelope could not produce any change in its level. Should, however, protrusion take place, as for example in the formation of a continental area, there would be a *sinking in the level of the water, and the volume of the waters which would recede from the shore lines would be exactly equal to the volume of land which appeared above the surface.* The newly created land surface would therefore owe its origin, first to the fact that it had been actually elevated, thereby increasing its distance from the centre of the globe of which it formed a part, and, secondly, to the fact that the waters had actually receded to fill a depression and had decreased their average distance from the centre.

The only escape from such a conclusion is the assumption that as continents have emerged from oceanic waters equal volumes of land have, at the same time, been subsiding beneath their surface. Not forgetting the arguments which have been brought forward to show that a Lemuria, an Antarctica, and an Atlantis or a Poseidonia may have sunk, together with the geological evidences of vast depressions and elevations, it seems unlikely that the conditions leading to the outlining of the existing continents should have been accompanied by the subsidence of land surfaces of equal volume.

Since, therefore, the bulk of the materials forming continents have been raised upwards from ocean beds, a fact testified by their stratified, fossiliferous, and other characters, the next object is to determine how far the position in which we now see them is due to actual

uplifting and to what extent it may be attributed to the
retreat of the waters. What we know definitely is that
the mean height of the continental areas relatively to
present sea level is something greater than 1,000 feet;
and that the relative areas of land and oceanic surfaces
are as 1 to 3. Assuming these numbers to be approximately
correct, if the land excrescences could be uniformly spread
over the bottom of the sea from whence they came, the
result would be equivalent to spreading a block of material
1,000 feet in height over an area three times as large as
that which it now occupies, while the waters would rise
to cover an area four times the size of that which they
now present. Neglecting the varying circumferences
theoretically involved in these operations, we calculate
that the waters would rise 250 feet above their present
level. With a mean height of land, as given by Dr. J.
Murray, of 1,937 feet, the apparent uplift due to the
recession of the waters would be 487 feet. When land
surfaces have gone up, then the oceanic level must have
gone down, and during geological times these movements
and their converse have been oscillatory and in opposite
directions.

To gain some idea of the extent to which the retreat of
the ocean into growing oceanic depressions has accelerated
the exposure of strata, we will suppose a stage in the
Earth's history when it was an uncrumpled sphere covered
by a deep ocean. With a mean oceanic depth of 15,000
feet, and a mean height of our continents of 1,000 feet,
the total height of the continental protuberances is 16,000
feet, and if this 16,000 feet of material could be spread
over a sphere drawn through the present *mean* depth of
the waters, such a layer would be 4,000 feet in thickness.
The Rev. O. Fisher in a similar calculation takes his
datum line through the *greatest* depth of the ocean, or
about 9,000 feet lower than the one employed here. When
this quantity is added to the 4,000 feet of my calculation, the results representing the dimensions of the
uncrumpled sphere are in accordance. By such a process

we obtain approximate dimensions for a primitive lithosphere, and the present waters distributed over such a surface would have a depth of 11,250 feet.

After solidification of the crust we cannot imagine changes of any magnitude taking place in this crust due to its own contraction by further loss of heat. The only deformation it has suffered since it hardened has chiefly been in consequence of accommodating itself to a shrinking nucleus.

With conditions somewhat of this nature, we are in a position to sketch the general character of the changes which have succeeded each other in the relationship of land to water during the evolution of continental areas.

From the investigations of Dr. A. v. Tillo it appears that the relative areas of the different geological groups, as at present known, stand to each other in the following proportions :

Archæan	20·3
Palæozoic	17·5
Mesozoic	20·2
Tertiary, &c.	42·0

Although some 27 per cent. of the surfaces of the continents are unexplored, it is not likely that the relation between these numbers will be greatly altered. As the sum of the above numbers represents the present land area, which is one-third of the oceanic area, then we can approximately determine the ratio of land to sea at the termination of each of the preceding epochs.

The values of land to sea would be as follows :

Archæan	1 : 19
Palæozoic	1 : 10
Mesozoic	1 : 6
Tertiary, &c.	1 : 3

The next factors required are a series of numbers representing the mean heights of successive land areas. If these are assumed to be proportional to the thickness of the rocks which constitute them—the figures representing which are, according to the investigations of Dr. Haughton, proportional to the time taken to form such strata—the following table is obtained :

	Feet
Archæan	343
Palæozoic	768
Mesozoic	913
Tertiary, &c.	1,000

If, however, the mean heights are proportional to the land areas exposed, the table becomes:

	Feet
Archæan	203
Palæozoic	378
Mesozoic	580
Tertiary, &c.	1,000

From what we know of the growth of mountain ranges, which have added largely to the height of continental areas, especially in Tertiary times, and because it seems likely that a great increase in land area means a correspondingly large increase in average height, the latter table is the one which will be employed.

It will be observed that the possible inaccuracies in the foregoing data will depend upon the ratios which have been assumed respecting the relation of land to oceanic area at the close of certain epochs in geological history. The last of these ratios, because it has been determined by actual measurement, cannot be far from the truth, but the remainder are more and more uncertain as we proceed back in time. Notwithstanding these inaccuracies, and admitting that they are extremely large, it seems impossible that these data should fail to lead us to a truer idea of the changes which have taken place in continental development than sheer guesses would furnish. The commencement of the evolutionary process we wish to trace may be taken at the end of Archæan times, when by the deformation of a primitive sphere buried beneath 11,250 feet of water a continental area has been exposed, the area of which, relative to that of the surrounding waters, is as 1 : 19, while its mean height is 203 feet. To bring this about, there must have been a real elevation of 193 feet, while 10 feet more has been exposed by a vertical fall in the waters receding to occupy the depression formed by the uplifting of the land. The fall in the mean depth of the ocean would be 602 feet, and its mean depth would increase to 11,842 feet. The general slope of the land along a line 3,000 miles (18×10^6 feet) in length, which may be taken to represent an average slope between the centre of a continental area and the bottom of the surrounding ocean, would be such that the 10 feet of vertical fall would expose a fringe of land along the coast with an average breadth of 11,623 feet.

If we treat the other cases similarly and tabulate the results, the relationship of land to water at the termination of successive geological epochs may have been something like the following:

—	Mean height above sea level	Vertical exposure due to fall in water	Exposure due to actual elevation	Mean depth of ocean	Depression in ocean bed	Breadth of shore exposed by fall in water
Archæan	203	10	193	11,842	602	11,623
Palæozoic	378	34	344	12,375	1,159	47,988
Mesozoic	580	50	497	13,125	1,958	109,011
Tertiary	1,000	250	750	15,000	4,000	281,250

As there is a great want of exactness in the data on which the above table is founded, it can only be looked at as suggestive of the character of the changes which have brought about the present relationship of land and water. The average breadth of the existing shore lines due to the retreat of the ocean is seen to be about 47 miles, but had the average slope been measured along a line 6,000 miles in length rather than 3,000 miles this quantity would have been doubled.

A general conclusion at which we arrive is that elevations due to bradyseismical movements have a magnitude which, when measured vertically, is 25 per cent. less than that usually attributed to them.

Should we only desire an estimate of the superficial area of land which may owe its existence to these movements, we may look at the fractional portion of continents which would remain if the present sea level were raised 250 or 300 feet.

For the geographer who gives attention to the evolution of the superficial features of our globe oceanic metahypsosis is a factor that has played an important, but often neglected rôle.

The next and last section of this chapter is to show that although, on the whole, there has been during geological time a considerable fall in ocean level, the process has been oscillatory. When any considerable area of the Earth's crust commences to yield under the influence of those forces causing elevation or depression, it is easy to imagine that a movement once started should continue in its initial direction, but it is difficult to picture conditions which should result in a reversal in the direction of such bradyseismic operations. Nevertheless, there are many geological arrangements of strata—when, for example, land surfaces are buried beneath marine deposits of considerable thickness—which apparently demand such reversals. The question which arises, however, is whether it is always necessary to avail ourselves of an unlimited bradyseismical credit when explaining the stratigraphical history of our earth.

To answer the query we will consider what happens on an area of elevation as it gradually emerges from its ocean covering, and then when the same is buckled, as in the process of mountain formation.

Round the shore line of such a region of elevation a series of strata will be deposited one above the other, so that after their emergence we can walk seawards from an inland primitive nucleus across the outcrops of successively newer strata. Such a succession is well exemplified in the general arrangement of the Palæozoic series of the North American Continent.

Whilst this process is in operation large areas of land will also be appearing on shallow shores by the withdrawal of the water.

The next step is to assume that on the rising dome buckling or folding takes place, as, for example, during the formation of the Urals and other mountains towards the close of Palæozoic times, and again in early Tertiary times, when the Himalayas, the Alps, and many of the largest of existing European ranges were slowly brought into existence.

As these mountains were elevated—which, as indicated by the complexity of their folds and the contorted and faulted strata of which they consist, took place spasmodically —it is likely that to their right and left equal volumetric depressions of the land were formed. Because the downward motion was intermittent, it is further probable that by sedimentation the sinking surfaces would be restored and the series of strata deposited would alternate in their character like those met with in the coal measures.

Around the world, in regions where there were no movements of the Earth's crust along shallow shores, a fall in the ocean level would be marked. On areas slowly subsiding large tracts might be exposed to support vegetation, which would subsequently be covered by marine deposits.

Giving the Carpathians and the Himalayas breadths of 180 and 240 miles respectively, and the remaining Tertiary mountains breadths of 60 miles, their lengths being what we see on an atlas, these elevations cover an area of about 500,000 square miles. If ranges parallel to the Pyrenees, the Dinaric Alps, the Balkans and their continuation into Turkey, ranges parallel to the Himalaya, the Andes and other mountains were uplifted about the same period, the elevation area may easily have extended over 1,000,000 square miles.

If the mean height of upheaval in these mountainous regions was about 4,000 feet, which is the present mean height of Switzerland, and if we know the ratio of land to water about this time, we are enabled to form a rough idea of the amount of general depression in oceanic waters which accompanied the uplifting of the Tertiary mountains. From what has already been said it would seem that the required ratio was between 1 : 6 and 1 : 3. With a ratio of 1 : 4 the total amount of vertical fall in ocean level which took place step by step would be about twenty-six feet.

Without insisting that all evidences of subsidence are to be explained as the result of the formation of secondary features on the Earth's surface, it is at least remarkable that the two great mountain-forming epochs have a close chronological identity with the two great periods of coal formation.[1]

If this relationship between marked exhibition of bradyseismical activity and the formation of coal is not admitted, we are still at a loss to explain why those conditions which led to the formation of coal were only marked at two particular epochs in geological history, and why they were exhibited simultaneously at so many points round our globe.

---

[1] There is also a close relationship between the periods of mountain formation and volcanic activity.

CHAPTER II

METHODS OF MEASURING BRADYSEISMICAL MOTION

Geological measurements of contraction—Vertical and horizontal changes
in the relative position of two points and changes in the inclination
of a line joining the same—Measurement of elevation on the Baltic
coast—Measurement of differences of elevation of two points relatively to water level—The Potsdam water level—The use of horizontal pendulums and spirit levels—Measurements relatively to
anticlinal folding.

FROM what has been said respecting the growth of
continents and changes accompanying the formation of
mountain ranges, it will be evident that the measurement
of these movements by reference to sea level may lead to
results which are misleading. In bradyseismical changes
which have taken place in recent times, or within the
limits of historical periods, such a datum may be all
sufficient. All that we know definitely about these
movements is that after a long period of years, or after
an unmeasurable geological interval, certain results have
been brought about, and for further information respecting
the greatest of all movements in the crust of our earth
we are driven to speculations based on observations which
are more characterised by their number and similarity
than by their accuracy and variety. The most important
of these relate to the vertical displacement of a seaboard
relatively to the surface of a neighbouring sea or ocean,
from which we infer that in certain places such movements may have had rates of from a quarter to one inch
per annum.

To extend our knowledge it seems reasonable that

we should first consider the probable character of the phenomena we expect to find, having done which, methods of investigation likely to lead to good results may then suggest themselves.

It is generally admitted that strata which were once practically horizontal have, by tangential thrust or other causes, been slowly deformed until they have assumed a ridge or wave-like form. To study the movements by which this change has been brought about there are three important points to which attention may be given, namely, the horizontal, vertical, and angular movements.

Thus Prof. A. Favre estimates that strata forming certain mountains in Savoy have been compressed one-third.

Prof. Claypole estimates that 100 miles run on the Appalachians have been brought within the space of 65 miles.

Prof. T. C. Mendenhall, formerly chief of the United States Coast and Geodetic Survey, taking the distance between the Atlantic and the Pacific at sea level on the 39th parallel as unity, finds that the ratio of this to the actual profile across the Alleghany region is 1·00096, while for the Rocky Mountain region it is 1·00147.

The coming together of points during elevation should be most marked along lines at right angles to the axis of elevation.

Vertical motion, again, we might expect to be most marked near to the crest and sinus of a fold, at the former the direction of the motion being upwards and at the latter possibly downwards.

Between these loci the vertical displacement will vary.

Finally, throughout the district of movement, excepting on the axis of elevation and depression and on lines parallel to their axis, there has probably been a varying change in inclination to the horizon.

For the full determination of the changing configuration of an area, these three elements should be studied, but nearly all that has hitherto been done for the experimental demonstration of these changes has had to do with the horizontal movement only. After the great Japan earthquake of 1891, it was painfully evident that the horizontal distance between the foundations for the piers of bridges had been shortened, river beds had been contracted from 1 to 2 per cent. of their former width, with the result that floods were anticipated, while for plots of ground which had been reduced in length in the ratio of 10 : 7 re-surveys were required for assessment.

Had this same result been attained at an ordinary secular rate, it would probably have been far too slow to have been brought within the reach of experimental demonstration.

At first sight it might be thought that changes in the horizontal distance between two points taken, for example, on a line at right angles to an axis of mountain folding might be determined by measurements after sufficiently long intervals of time of a base line carried in such a direction. Inasmuch as the variation on such a length would only change with the versed sine of the angular tilting, the method is not one that would be likely to yield results of any value.

The often published results relating to alterations in the level of the Baltic relatively to certain markings which in the early part of last century were cut upon the rocks are well known to all students of geology.

This method of measurement is direct and simple, but it only tells us whether a change has taken place after a long period of years.

If the change is measurable in feet we are content to accept as a fact that there has been an upward or downward movement of approximately that extent; but if it has been a change of a few inches, so many causes by which the water level might be slightly influenced suggest themselves, that we find difficulty in reconciling ourselves to the conclusion that the movement has really been that of the land.

For example, great rivers pouring a variable amount of sediment and water into a sea like the Baltic, fluctuating winds, varying barometric pressure, alteration in the depth of channels leading to the open ocean, and other causes, tend to alterations in the mean level of the datum, For reasons like these the idea suggests itself that by endeavouring to measure the *relative* elevation of two or more points on a shore line, rather than attempting to measure the absolute elevation of any one, we might perhaps obtain more trustworthy results. To carry this out we might, for example, observe the difference in the records obtained at the same time from two or more tide gauges situated along a coast where the rise and fall of tide were not excessive.

If there is no change taking place in the sea level relatively to the land, then these differences between the heights measured at the various stations, which heights are measured relatively to certain bench marks at those stations, should when the tide is in the same phase and there are no disturbances—for example, the piling up of the water by the wind—remain constant. The chief assumption here made is that during similar phases of the tide, the surface of the water has the same configuration. By means of a system of stations in nearly a straight line, the configuration of the water surface under varying conditions along that line might be determined. As an illustration of how the work might be systematically performed we will assume that we have at least three tide gauges, from ten to twenty miles apart, installed round the shores of Tokyo Bay. With this installation on a series of consecutive days, we can readily determine the following particulars :

1. Total rise of water from low water to high water.

2. Whether the tide is increasing or diminishing from day to day.

3. Whether at any point in the vicinity of the mouth of the bay there is no tide.

We can then for one or all of these days determine the height of certain bench marks relative to high or low water (these being convenient phases of the tide), or the height at each of the stations above water level at the same time.

Again, say a year afterwards, let us make similar observations when the tide has the same total rise and is increasing or diminishing, as in the previous year. Also, it will be necessary to determine whether the point of no-tide has remained fixed, and, if not, re-determine the water configuration. Then the difference of the differences between the indications at the several stations on the one occasion and those on the other occasion will measure the relative rise or fall.

Any difference in the height at any one station is an indication of total rise. It is evident that in order to measure these changes and to determine the axis of the movements it is necessary to make observations at at least three stations.

In reply to the query as to the amount of change we expect to measure, we may say that the evidences of elevation round the Bay of Tokyo are sufficient to lead us to expect changes at least equal to that which, for example, has been determined on the coast of Italy.

If we wish to dispense with inaccuracies due to fluctuations in sea level, our only recourse appears to be the process of ordinary levelling and redetermining at stated intervals the difference in height along a line at right angles to an anticlinal fold.

Col. J. Farquharson, C.B., R.E., tells me that the average error of seven men levelling along roads, over distances of between fifty and sixty miles, is ·312 inch per mile.

Some years ago it was suggested in Japan that new changes in elevation might possibly be noted and measured by means of a long water level, the direction of this level being at right angles to an axis of elevation.

The details of the arrangement and its feasibility were discussed at some length, but, partly because it was recognised that very similar results might be obtained by the proper installation of a few horizontal pendulums, and partly on account of the difficulties and expense in constructing such a water level, no attempt was made to carry out the suggestion.

In connection with the Geodetic Observatory at Potsdam, Dr. Kühnen has arranged four water levels, each 200 metres in length, which are constructed to form the side of a square. By means of special arrangements at each corner of the square the height of the water at that corner can be determined.

Lastly, we turn to the observations which may be made on changes of level at one or more points. The reason why instruments like horizontal pendulums rather than instruments like astronomical levels seem to be more suitable for investigations of this description is discussed in another chapter.

In Japan both classes of these instruments have been used for long periods of time, and, as in other countries, it was observed that both the pendulums and the bubbles of the levels moved irregularly. With the pendulums it often happens that a diurnal period is observed, which exists as a wave superimposed upon a more general movement. In Europe these instruments seem always to have been placed to record motions at right angles or parallel to the meridian. In Japan for many years a similar course was adopted, but for some years the installations were such that the movements recorded have been parallel, or at right angles, to the axis of an anticlinal fold. The result of three years' observations showed that there was a gradual tilting towards the west (see ' Brit. Assoc. Reports,' 1896).

CHAPTER III

CAUSES OF EARTHQUAKES

The views of Aristotle, Pliny, the Chinese, Shakespeare respecting the
cause of earthquakes—Myths relating to subterranean animals—The
Scandinavian Loki—Earthquakes due to human wickedness—
Electrical theories—Seismo-chemical theories—Earthquakes due to
volcanic action—The distribution of seismic activity shows that
earthquakes are frequent in regions of bradyseismic action—The
earthquakes of the Himalaya, Switzerland, Japan, and along the
steeper flexures of the earth's crust—Submarine disturbances—The
greater number of earthquakes are due to fracturing of the earth's
crust or the movements of a quasi-elastic magma.

THE genus *terremoto* has its species, all of which, even
when they fail to create alarm, arouse our curiosity as to
their origin, a subject about which the world has specu-
lated throughout all ages. With our present knowledge
respecting changes which are in operation in and beneath
the crust on which we live, we have not to go far to find
causes which, singly or in conjunction, are amply sufficient
to shake the ground. The greatest difficulty which
presents itself is to select from the causes which may
possibly produce earthquakes those which play the most
important part in the creation of seismic sensibility, and
at the same time not to confound them with minor
influences which may cause a region in a state of seismic
stress to suddenly collapse. In the present chapter there
is no intention to try and deal with gravitational effects
of the sun or moon, or with the effects of barometrical or
other loads—the stresses due to which may result in
yieldings being more frequent at one season than at
another—but only to refer to causes which bring about

conditions to which earthquakes are more directly attributable.

As an introduction to the modern views respecting the causes of earthquakes, it will be not without interest to recapitulate briefly the opinions which have been held in the past. In early times, earthquakes, displays of volcanic activity, the fossils buried in the rocks, and other things which to the savage have always been unintelligible, were by a few philosophers attributed to natural causes. In the middle ages the teachings respecting such phenomena were that their explanation was only to be found by an appeal to the supernatural, and it was not until the eighteenth century that the educated world, armed with the results of observation, returned to the doctrines of the ancients. Aristotle, Pliny, and other philosophers, whose writings testify to the fact that they had observed steam and other exhalations escaping from volcanic vents, held that earthquakes were due to the working of wind or imprisoned vapour beneath the earth's crust—a view which finds its parallel in the early philosophy of the Chinese. Natural theories of this order are to be met with until late in the middle ages. Shakespeare in his ' Henry IV.' says—

> Diseasèd nature oftentimes breaks forth
> In strange eruptions : oft the teeming earth
> Is with a kind of colic pinch'd and vex'd
> By the imprisoning of unruly wind
> Within her womb ; which, for enlargement striving,
> Shakes the old beldam earth, and topples down
> Steeples and moss-grown towers.

Co-existent with these doctrines, which are yet to be found amongst the uneducated, are the superstitions that earth shakings are due to the movement of a subterranean god or some mythical monster. In Japan, for example, it was supposed that there existed beneath the ground a large earth spider or ' *jishin mushi*,' which later in history became a cat-fish. At Kashima, some sixty miles north-east from Tokyo, there is a rock which is said to rest upon the head of this creature and keep it quiet. At this place,

therefore, earthquakes should not be frequent. The rest
of the empire is shaken by the wriggling of its tail and
body. In Mongolia the earth shaker is a subterranean
hog; in India it is a mole; the Mussulmans picture it an
elephant; in the Celebes there is a world-supporting hog;
while in North America the subterranean creature is a
tortoise. The people of Kamtchatka had a god called
Tuil, who, like themselves, lived amongst the ice and
snow, and when he wanted exercise went out with his
dogs. These dogs were, it was supposed, infested by
insects, and when now and then they stopped to scratch
themselves, their movements produced the shakings called
earthquakes. In Scandinavia, which is essentially the land
of mythology, there was an evil genius named Loki, who,
having killed his brother Baldwin, was bound to a rock,
face upwards, so that the poison of a serpent should drop
on his face. Loki's wife, however, intercepted the poison
in a vessel, and it was only when she had to go away to
empty the dish that a few drops reached the prostrate
deity and caused him to writhe in agony and shake the
earth. As other illustrations of the stimulating effects
which seismic and volcanic activities have at all times
exerted on the mind, we need only mention Pluto, Vulcan,
and Poseidon, whilst the command that we are not to
make the likeness of anything that is in the earth beneath
suggests that in the time of Moses a subterranean mythology
existed which barred the way to religious progress.

In consequence of numerous shocks, which in 1750 were
felt throughout Great Britain and were followed five years
later by the terrible catastrophe which overtook Lisbon,
and because of the general activity of seismic and volcanic
agencies which about this time made itself manifest
throughout the world, universal interest was attracted to
earthquake phenomena. Many of the theories which were
then propounded to explain the origin of these mysterious
occurrences are embodied in sermons, the authors of which
tell us that earthquakes are direct visitations from above,
brought about by man's increasing wickedness. In a

pamphlet about the earthquake at Palermo in 1706, we read that ' the people seemed to be extremely humble and penitent, scourging themselves and doing penance,' and in conclusion there is the remark that ' it was generally apprehended that this was a mark of God's vengeance for the immorality of the inhabitants.' The ideas then prevalent are summed up in a little poem called ' The Earthquake ' written in 1750. It runs as follows :

> What pow'rful hand with force unknown,
> Can these repeated tremblings make ?
> Or do th' imprison'd vapours groan ?
> Or do the shores with fabled Tridents shake ?
> Ah no ! the tread of impious feet,
> The conscious earth impatient bears ;
> And shudd'ring with the guilty weight,
> One common grave for her bad race prepares.

The views set forth in the last four lines of this poem still find expression from time to time. After the earthquake which in 1883 alarmed the inhabitants in Charleston, the negro preachers told their congregations that the disturbance had visited that city in particular in consequence of its sins.

Again, in 1891, after the great earthquake which devastated Central Japan, evidence of a selective providence was found in the fact that a few of the houses tenanted by Christian converts happened to remain standing amongst the ruins of their Buddhist and Shinto neighbours.

A theory respecting the cause of earthquakes, the reasons for which have hardly yet been given by its most ardent advocates, is that these phenomena are the result of electrical discharges.

As an indication of the popularity which the electrical theory of earthquakes has had, I give a list of a few of its more distinguished advocates. In Italy, Beccaria (1753), Della Torre (1777), Sarti (1783), Mignani (1784), Vivenzio (1788), Cavallo (1790) Fellini (1791), Poli (1783), Toaldo (1798), Vassalli (1808), Matteucci (1829), Sanna Solaro (1887). In other countries we find Priestley (1762), Monteyro (1765), Buffon (1781), Bertholon (1787), Brisson

(1803), Patin (1820), De Bylant, &c. The hypothesis that electricity, by causing the explosion of subterranean gases, has indirectly resulted in earthquakes, has been put forward by Olivi, Boccardo, and Bombicci (' Bollettino della Società Geologica Italiana,' vol. ix. fasc. 1 ; ' Fenomeni Elettrici Magnetici dei Terremoti,' Mario Baratta).

The first suggestion that there might be a relationship between the actions which so violently disturb the atmosphere and those which shake the earth, I find in a quotation given by Baratta from Io. B. Portae ' De aeris transmutationibus,' 1614, who says :

' Nihil aliud terraemotus est quam subterraneum tonitruum, et tonitruum est coelestis terraemotus.'

The possibility that earthquakes may in any way be connected with electrical phenomena is discussed in Chapter XII.

Seismo-chemical theories seem to have had their origin with Vannuccio Biringuccio, who about 1550 wrote ten volumes entitled ' Pirotechina,' in which he advanced the idea that earthquake motion was due to some subterranean explosion. Those who, following him, adopted the same idea, endeavoured not only to define the nature of the materials employed in the operation, but the conditions under which they were accumulated in caverns and the method by which they were ignited. Although bitumen and sulphur were thought to have played an important part in the production of explosive gases, the *materia pinguis* or ' fatty matter ' of Agricola, which by its fermentation gave birth to fossils, was called upon by no less an authority than Des Cartes to produce by similar processes a ' fatty vapour ' which by its ignition and explosion shook the earth.

The action of water upon quick lime was not neglected, while iron pyrites, as a material yielding sulphurous vapours, was a substance that found favour with many writers. Even as late as 1683 Lyster suggested that earthquakes were more frequent in Italy than in England because the pyrites of the former country might be richer in sulphur than that of the latter, while caverns in which the gases might accumulate were probably most numerous in the most frequently shaken districts.

The ignition of the various gases was attributed to fermentation causing spontaneous combustion, the friction and impact of falling rocks, the heat developed by combination, and to other causes. Although Lemery in 1703 with a mixture of iron filings, sulphur, and water succeeded in producing the appearances of a volcanic eruption, and like Gassendi, who preceded him, may be accredited with having appealed to experiment to support his views, a little knowledge of chemistry, like a little knowledge of electricity, did much in misdirecting inquiry from its true course.

About the middle of the eighteenth century the idea that earthquakes might in some way or other be connected with volcanic action was revived. Michell, writing in 1760, observes that earthquakes chiefly occur in volcanic countries, and suggests that they are the immediate result of steam forcing its way between stratified accumulations in the endeavour to establish an active vent. This view is modified by Rogers, who attributes the pulsatory motion of the surface to the passage of molten lava between the planes of bedding of subjacent rocks.

From this time up to the present many earthquakes have with good reason been attributed to volcanic action. Humboldt tells us in general but vague language that earthquakes and volcanoes result from a common cause, which is 'the reaction of the fiery interior of the earth upon its rigid crust.' Mallet, who devoted so much time to the study of subterranean phenomena, shows at great length that in all probability earthquakes are due to the sudden evolution and condensation of steam, the accompanying explosion, which may be repeated, often resulting in the production of faults and fissures. In the concluding chapters of his classical work on the Neapolitan earthquake of 1857 he shows the effect of water entering heated cavities, where it assumes the spheroidal state and is superheated. On the cessation of these conditions, instantaneous evaporation takes place, accompanied by violent explosion. In 1890 a theory similar to this is discussed by M. Baratta ('Bollettino della Società Geologica Italiana,' vol. ix. fasc. 2).

We know from observation that before a volcano bursts into eruption there may be many ineffectual efforts to establish a vent, and that each of these is announced by a sudden shaking of the ground. The final and successful effort is usually accompanied by movements more pronounced, and from these observations alone it is reasonable to suppose that at least certain earthquakes are the immediate outcome of subterranean volcanic action. Should the effort be unusually large, resulting in the disappearance of half an island or a large mountain, as was the case in 1883 at Krakatoa and in 1888 at Bandaisan (Japan), the earth shaking is correspondingly greater.

Although it is admitted that whenever effects of this description are manifested on the surface, much of the initial energy has been expended in projection, it is remarkable that the accompanying earth shaking has been perceptible over a comparatively limited area. For example, the area shaken at the time of the Bandaisan explosion was less than 2,000 square miles. If we compare figures like these with those which represent earthquakes, some of which originate in non-volcanic districts, and which are repeated many times per year, they are insignificantly small.

To produce earthquakes which are felt over areas of five or ten thousand square miles, and which give rise to waves which may be recorded at any point upon our globe, it is difficult to imagine how the primary impulse could have originated at a volcanic focus. Volcanic explosions, as we see them, seem to result from the concentration of subterranean energy at a point, while to shake the whole surface of our globe it would appear necessary that the initial effort should be exerted on a surface very much larger than we can reasonably suppose to exist beneath a volcano.

A very much more serious objection to the volcanic origin of the majority of earthquakes is the fact that these disturbances are common in the Himalaya, Switzerland, and other non-volcanic regions. The destructive earthquake in 1891 in Mino and Owari occurred in a region of metamorphic and stratified rocks. Again, an analysis of some 10,000 earthquake observations in Japan shows that there have been but comparatively few which had their origin near to the volcanoes in the country. The greater number of this series originated beneath the ocean or along the seaboard, and as they radiated inland they became more and more feeble, until, on reaching the backbone of the country, which is drilled by numerous volcanic vents, they were almost imperceptible. Beyond this central range of mountains, earthquakes are only rarely experienced, and what is true for Japan seems to be generally true for the coasts of North and South America.

Throughout the world we find that seismic energy is most marked along the steeper flexures in the earth's crust, in localities where there is evidence of secular movement, and in mountains which are geologically new and where we have no reason for supposing that bradyseismic movements have yet ceased.

As examples of the flexures to which reference is here made, we may take sections running at right angles to the coast lines of the various continents. The unit of distance over which such slopes have been measured is taken at 2 degrees, or 120 geographical miles. The following are a few of such slopes :

West coast, South America, near Aconcagua	. 1 in 20·2	
The Kurils from Urup	. 1 in 22·1	Seismic
Japan, west coast of Nipon . . .	. 1 in 30·4	districts
Sandwich Islands, northwards . . .	. 1 in 23·5	
Australia generally	. 1 in 91	Non-
Scotland from Ben Nevis	. 1 in 158	seismic
South Norway	. 1 in 73	districts
South America, eastwards	. 1 in 243	

The conclusion derived from this is that if we find slopes of considerable length extending downwards beneath the ocean steeper than 1 in 35, at such places submarine earthquakes, with their accompanying landslips, may be expected. On the summit of these slopes, whether they terminate in a plateau or as a range of mountains, volcanic action is frequent, whilst the earthquakes originate on the lower portions of the face and base of these declivities.[1] Districts where earthquakes, often followed by submarine disturbances, are most frequent are regions like the north-east portion of Japan and the South American coast between Valparaiso and Iquique. Here we have a double folding. The sea bed as it approaches the shore line, instead of rising gradually, sinks downwards to form a trough parallel to the coast, after which it rises to culminate in mountain ranges. The South American trough, which lies within fifty or sixty miles of the coast, like the Tascarora deep off Japan, attains depths of over 4,000 fathoms, and the bottoms of these double folds are well known origins of earthquakes and sea waves.

If we turn from these general illustrations and examine the conditions accompanying seismic activity, for example, in the Alps, the Himalaya, the Andes, or in the Peninsula of Italy, we find that we are in a region where mountain formation is geologically of recent origin, and where there is no reason to believe that the forces which brought these mighty folds into existence have yet ceased to act. In Italy and Japan, where there is a datum like sea level to which we can appeal, we learn that secular movements are yet active. From the maps of Taramelli and Baratta, showing the past and present distribution of seismic activity in Italy, it is evident that the greater number and the most severe disturbances follow the backbone of the peninsula. A map of these meizoseismic areas taken by themselves would fairly well represent the string of Miocene islands round which the remainder of the country has been built.

[1] See note upon the ' Geographical Distribution of Volcanoes,' by J. Milne, *Geological Magazine*, April 1880.

In Pliocene times these islands became united and the Apennines were completed as a range of hills.

From this time the growth of the country was rapid, and, if we except a strip along the western coast from Leghorn to Naples, in Quaternary times Italy was as we now see it.

The clearly marked and comparatively rapid brady-seismical movements which during historical times have taken place along the shore line of the latest addition to the Italian kingdom are well known to all geologists. The conclusion to which such observations lead is that wherever we find in progress those secular movements which result in the building up of countries or mountain ranges, there we should expect also to find a pronounced seismic activity. Thus, while admitting a few small earthquakes to be volcanic in their origin, we recognise the majority of these disturbances as the result of the sudden fracturing of the rocky crust under the influence of bending. The after-shocks which so frequently follow large earthquakes announce that the disturbed strata are gradually accommodating themselves to their new position.

On an anticlinal, the yielding, as in Italy, apparently takes place chiefly along the crest of the fold, while on a monoclinal flexure, as round a great portion of the Pacific, the fracturing seems most frequent along the region of maximum bending or greatest inflection.

That the bases of monoclines are tracts where faults are frequent has long been recognised by geologists, the former being, in the words of Sir Archibald Geikie, 'an incipient stage' of the latter. More distinct evidence of faulting being accompanied by earthquake motion is the fact that many large earthquakes have been accompanied by faults which are visible on the surface. The terrible shock which in 1891 laid waste hundreds of square miles in Central Japan seems to have been the immediate result of a great fracture in the earth's crust which, according to Dr. B. Koto, can be traced for a distance of over sixty miles. The surface of the ground

D

on one side of this line has fallen some twenty feet below
its former level, but the maximum throw is in all probability
much greater than that which is accessible for direct
measurement. The main fault was accompanied by many
minor dislocations, horizontal displacements, and even
compression, so that a river bed has been narrowed, while
plots of ground which were originally forty-eight feet in
length have had this dimension reduced to thirty feet. In
the Neo Valley, where the devastation was greatest, whole
tracts of rice fields on one side of the fault were suddenly
lowered relatively to those on the other side, and on the
statement of peasants that after the earthquake the sun
appeared to rise earlier than it did before, we have evidence
that when one side of the bottom of the valley fell, the
bounding mountains fell with it. The horizontal and
vertical displacements which took place are evident to
every traveller through the district. A compression of
from 1 to 2 per cent. across the river beds had to be allowed
for by the engineers who reconstructed the fallen bridges,
while the remeasurement of land for Government assessment showed that certain areas had decreased in size. It
is no doubt difficult for those who live in districts where
convulsions like these are unknown to realise these state-
ments, but when they are admitted it is no longer difficult to suppose that such sudden changes could well have
taken place without serious displacements in the mountains
rising from the area where they happened. A tract of
country more than fifty miles in length which carried
mountain ranges several thousands of feet in height was
suddenly fissured along its length; accompanying this
there was a back spring of strata released from strain,
and a collapse by falling of a valley bottom and its bounding ridges. The magnitude of this impulse, received
almost simultaneously over a large area, caused Central
Japan to shake so violently that forests slipped down from
mountain sides to block up valleys, while earth waves were
created which travelled round the globe.

Here, as was the case with the Quetta earthquake in

1892, fracturing of the rocky crust of the globe and terrific
shakings have accompanied each other, the former with
its attendant phenomena being sufficiently adequate to
have produced the latter. Should it be contended that it
was the violence of the earthquake which produced the
faulting (and no doubt violent shakings may relieve areas
which are on the verge of yielding and thus be the cause
of secondary earthquakes), we seem compelled to admit
the existence of seismic strains of almost inconceivable
magnitude exerting themselves beneath non-volcanic
regions.

It is undoubtedly true that earthquake disturbances
are not generally accompanied by any visible fracturing
on the surface of the ground, but that they may be the
result of such fracturing is rendered probable by the fact
that they occur in regions where secular movements are
in progress, or at least where geological experience has
demonstrated that dislocations are numerous.

Disturbances originating beneath the sea, which are
much more numerous than those originating beneath the
land, likewise emanate from a region of strain. Mr. W.
G. Forster, who has paid so much attention to the earthquakes of the Mediterranean, tells us that they have been
accompanied by great subsidences of the sea bottom.
After the Filiatra shock in 1886 it was found, while
searching for a broken cable thirty miles off shore, that a
depth of 900 fathoms existed where previously there had
been only 700 fathoms, and that some four knots of the
cable were covered by the 'landslip.' Mr. Forster gives
several examples where cables have been broken at the
time of earthquakes, and he also shows that soundings
taken after shocks have been markedly different from those
taken before the shocks, and this even in non-volcanic
regions.

Another remarkable series of alterations in ocean depth
are those off the Esmeralda River on the coast of Ecuador.
Mr. M. H. Gray, of the telegraph works at Silvertown, tells
me that here cables have frequently been broken, and

during repairs soundings have been taken. From charts of these soundings it is seen that at places accurately fixed by bearings on the shore, depths have increased from 100 to nearly 200 fathoms. Although it is possible that cables might be interrupted and alterations produced in the configuration of a sea bottom as a result of volcanic action, it is usually supposed that they are due either to submarine landslips or submarine seismic action accompanied by landslips and faulting. As Mr. Gray points out, a sub-marine landslip may also be produced by the percolation of water from mountain ranges downward through inclined strata until it finds vent in the ocean bottom. The result is a weakening in the support of the overlying materials, which sooner or later slide down to a greater depth.

In ocean currents we see another cause tending to render steep slopes and overhanging shelves unstable. When these give way, the ocean depth may be changed, and if the mass of dislodged material is large waves may be produced. We do not, however, see that the sliding downwards of silt and rock, especially beneath water, would result in a shaking sufficient to be felt and ruin towns at a distance of many miles. Whenever cables have been broken at the time of an earthquake, which is not an uncommon occurrence,[1] submarine landslides, like those which on the land strip mountain sides of their forests and block up valleys, may have accompanied the submarine faulting.

Another and not impossible cause of earthquakes is based on the hypothesis that under the influence of gravity there are intermittent adjustments in the materials lying beneath the steeper flexures of the earth's surface. The distortions observed in fossils and pebbles, the difference in thickness of contorted strata, the *creep* in coal mines, and other geological phenomena, indicate that stratified materials constituting the earth's crust may flow, and it is therefore not unlikely that there may be a subterranean

---

[1] See 'Sub-Oceanic Changes,' J. Milne, *Geograph. Journ.*, August and September, 1897.

activity of this description around the steeply folded basal frontiers of continental domes. The idea involved is that there is no sharp demarcation between a contracting nucleus and an accommodating shell, and that the quasi rigid materials bulged upwards under horizontal pressures sink under the influence of gravity. Beyond the fact that the home of earthquakes is where we should expect movements due to such hypogenic activities to be pronounced, the only evidence we have which points to their real existence are the curious magnetic perturbations noted in or near to seismic regions. Prior to certain earthquakes in Japan magnetometers have been greatly disturbed, a possible explanation for which is that in their vicinity a magnetic magma was changing in stress, in temperature, or was actually in motion. After the earthquakes the needles returned to rest. As pointed out by Captain E. W. Creak, F.R.S., after the eruption of Krakatoa in 1883, a remarkable alteration in the amount and character of secular magnetic change is said to have been observed in Bombay, Batavia, and Hongkong, which might be a coincidence, or it might point to changes below the surface of our earth (see Chapter XII.).

Wherever bending is taking place in the Earth's crust we find earthquakes, while if this process is going on in the vicinity of an ocean we find both earthquakes and volcanoes. Although a volcanic explosion or an abortive attempt to establish a volcanic orifice has often caused the ground to shake, the greater number of disturbances are either due to rock fracturing or to equilibrium adjustments of a subterranean quasi rigid magma. The sudden eruption of a volcano may cause a local shaking or cause an area in seismic strain to yield. In this case the volcano is the parent of the earthquake. On the other hand, by the sudden shaking of the ground a vent which has been dormant for a long period of years may have its statical equilibrium destroyed; and the relationship is reversed.

For local shakings and mere tremors a volcano has proved itself a 'safety valve,' but how far volcanic eruptions have relieved pressure, thereby facilitating further yieldings which might culminate in earthquakes, has never yet been carefully investigated. We know that before an eruption the ground around the base of certain volcanoes has trembled, but that during the time of the eruption the shakings have been less pronounced, whilst when activity has ceased the movements of the ground have recommenced.

The general conclusions at which we arrive are that the majority of earthquakes, including all of any magnitude, are spasmodic accelerations in the secular folding or 'creep'[1] of rock masses; a certain number, particularly those originating off the mouths of large rivers like the Tonegawa in Japan, may result from the sudden yielding in the more or less horizontal flow of deeply seated material, the immediate cause of which is overloading by the deposition of sediments; whilst a few, which are comparatively feeble and shake limited areas, are due to explosions at volcanic foci.

[1] 'On the Horizontal Movements of Rocks, &c.' By William Barlow, Esq., F.G.S., *Quart. Journ. Geolog. Soc.*, Nov. 1888.

CHAPTER IV

SEISMOMETRY

Seismographs, seismometers, seismoscopes—Columns of various forms—
Projection seismometers—Fluid seismometers—Movements of water
in lakes, Seiches, and Rhussen—Nadirane of d'Abbadie—Wolf's
nadirane—Surfaces of mercury used by Mallet, Abbot—Levels—
Pendulums—Ordinary and bifilar pendulums—Pendulums as tromo-
meters—Darwin's bifilar pendulum—The long pendulums of
Agamennone, Vicentini, Cancani—Duplex pendulums of Gray,
Ewing, Milne—Horizontal pendulums as seismographs or gonio-
meters—The pendulums of Perrot, Zöllner, Close, Von Rebeur-
Paschwitz, Milne—Bracket seismographs of Ewing, Chaplin, Gray—
Rolling sphere and parallel motion seismographs of Verbeck, Gray,
West, Alexander, Ewing—Seismographs for vertical motion of
Wagener, Gray, Ewing, Milne –Apparatus to record tilting—Record-
ing surfaces—Time indicators—A seismograph used in Japan—
Microphones—The Perry tromometer.

THE principal object of the present chapter is to give a
general description of instruments which are used for
recording movements of the Earth's crust. Because these
movements are so varied in their character, some being
sharp and violent, others, although rapid, being so small
that they are unfelt, while a third class exhibit themselves
as long period changes in the vertical, it is evident that
the instrument which satisfactorily records one class of
movement may be altogether unsuitable for recording those
of another class.

Many of these instruments have had their origin in
Japan, and with them all, and with others besides, the
writer has had considerable practical experience.

The distinction which may be drawn between seismographs, seismometers, and seismoscopes is implied in their names.

A fourth class includes instruments intended to give information respecting tilting which occurs with certain earthquakes. These may be described as goniometers.

It is convenient to distinguish clearly between horizontal pendulums, which are designed to follow changes in the vertical, and instruments in which a heavy mass remains practically at rest during an earthquake. These are accordingly called bracket seismographs. Instruments which record minute movements of the soil, which may be true elastic vibrations, are known as tromometers or tremor recorders.

Columns

It has often been suggested that something about an earthquake might be learnt from its overturning effect upon a-column standing freely on its base.

In Japan experiments were tried with small columns which were square, cylindrical, conical, and of other forms, but the results obtained were almost valueless. It often happened that a series of columns standing on equal bases fell in all azimuths, and that sometimes the columns with large bases fell, while those with smaller bases, on which they stood with difficulty, remained upright.

When it is remembered that a column before falling tends to oscillate with a period varying with its amplitude of swing, and that while oscillating it has a tendency to rotate, while at the same time the Earth's motion may vary in amplitude, period and direction, the reason that column seismometers have proved unsatisfactory is apparent.

Large heavy columns, like gravestones, which are not affected by tremors and can only be overturned by pronounced shocks, have often furnished valuable information about the maximum accelerations and the direction of movements which have overturned them. If rotation has taken place, certain inferences may also be drawn as to the nature of the earth's movements.

When the period of motion is short we should expect a body to fall inwards or towards the origin of motion. On the other hand, when the period is long the body may

FIG. 1.—OVERTURNED GRAVESTONES, SHONAI, 1893 (ŌMORI)

move with the ground, and acquire its velocity to fall outwards or away from the origin at or about the time the ground commences to swing backwards.

A sensitive seismoscope may be made with pins or thin strips of glass, which are unable to stand by themselves, but are propped against a suitable support.

If it is simply desired to make a body fall in consequence of a slight mechanical shaking, the body may be balanced on the top of a pointer projecting upwards from the segment of a sphere heavily loaded near to its centre of oscillation (fig. 2). By the fall of such a body a catch connected with apparatus intended to set a record-receiving surface in motion may be released.

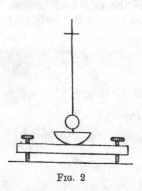

Fig. 2

Projection Seismometers

For many years there has been standing outside the Seismological Laboratory at the University of Tokyo a post, round the top of which is a horizontal ledge, carrying a set of iron balls. When it was put up it was expected that at the time of a strong earthquake these balls would be projected, and that, from the one projected the farthest, the direction of principal motion and its maximum velocity might be learned. Although Tokyo has suffered many severe shakings, the most that has happened is that one or two of the balls have fallen at the foot of the post. If earthquakes commenced with a single decided shock, it is possible that this contrivance, which is of very ancient date, might possess certain merits; but as this is not the case, the device yields nothing of any value, and under no circumstances could it yield anything better than would be furnished by the projection of tiles, coping-stones, and similar objects common to all cities.

Fluid Seismometers

A writer in the 'Quarterly Review' (vol. lxiii. p. 61) has criticised the use of vessels filled with fluid, the wash of which can be recorded by the mark which it leaves upon the bounding walls, as 'ridiculous and utterly impracticable,' but it must not be forgotten that the direction in which an earth movement has taken place has often been recorded by contrivances of this nature. In Tokyo for many years the only records of earthquake direction depended upon the indications given by a portion of Palmieri's seismograph, which consists of horizontal glass tubes turned up at either end and partly filled with mercury. These tubes pointed in different azimuths. A comparison with other instrumental records shows that the tube in which the greatest wash took place indicated fairly well the direction of principal motion. The general principle of this instrument is very similar to one described by the late Robert Mallet in 1846 ('Trans. R.L.A.' xxi. p. 107), in connection with which he gives the following formula.

If τ is the period in seconds of the moving fluid, l its depth measured in feet, and $g=32\cdot2$ feet per second, per second then

$$\tau = \pi \sqrt{\cdot \frac{5 \times l}{g}}.$$

Because the natural period of a fluid in an arrangement of this kind depends upon the square root of its depth, and there is variability in the period of an earthquake, the records for intensity as determined from the height of the wash are unreliable.

In the geodynamic levels of Professor G. Grablovitz, which are 2·5 metres in length, terminating in vessels 25 cm. in height filled with water, the movement of the fluid by means of a float actuating a lever, the other end of which writes on the smoked surface of a drum, is magnified fifty times. Whilst they have recorded certain seismic movements, they apparently failed to record the long period pulsations of the Japan earthquake of June 15, 1896 (see 'Seismological Investigations,' in 'British Association Reports,' 1896).

When we read that in a certain district the earthquake motion caused water to be thrown out of buckets, or the fluid from the vats of dyers, which is a common way or describing the intensity of a disturbance in Japan, we know from experience that the severity of the movement has been sufficient to dislodge tiles and shatter ordinary brick chimneys. If in a tank or pond we see the water rushing across its length or breadth, rising a foot or two on one side and then a foot or two on the other—as was the case in Tokyo on October 28, 1891—we know from this observation, taken in conjunction with others, that long undulations are passing beneath our feet, which are the form which shorter period and more violent movements in an epifocal area assume after having travelled a considerable distance.

In 1755, after the great earthquake in Lisbon, motion of this description disturbed the meres in England and the great lakes of North America. Whether any of the Seiches and Rhussen of the Swiss lakes, observed and studied by Professor F. A. Forel, can be traced to a similar origin is doubtful.

The formula connecting the periodic movement of a large body of water and the dimensions of the containing vessel is $\tau = \dfrac{l}{\sqrt{gh}}$,

where τ = the time of oscillation in seconds,

l = the length of the vessel in the direction of oscillation,

h = the mean depth of the water in units similar to l,

and where g is also expressed in the same units.

From this it follows that $h = \dfrac{l}{g\tau^2}$ or $\dfrac{v^2}{g}$.

The latter formula, known as Russell's, may be used in making approximate determinations of the mean depth of an ocean, across which an earthquake wave has been transmitted with a known velocity.

Although contrivances in which the records of an earthquake depend upon the violent disturbance of a quantity of liquid give but little information of value about earthquakes, the observations of changes in a reflecting surface of mercury due to actual tilting or to elastic vibrations have led to results of considerable importance.

In the nadirane of the late M. d'Abbadie, which has been established for many years at Abbadia in the south of France, the pool of mercury is in a well 6 ft. 6 in. deep excavated in the solid rock. Above this there is a cone of concrete 26 ft. in height pierced down the centre by a hole, at the top of which cross wires are arranged so that by means of a microscope these and their reflections from the mercury can be seen simultaneously. The apparatus is capable of measuring a change of angle of $0''\cdot03$, and it is therefore of value in measuring displacements in the vertical. The fact that the images are seldom at rest indicates that the apparatus behaves as a tromometer.

The nadirane of M. Wolf, designed for use at the Paris Observatory, chiefly differs from that of M. d'Abbadie in the following particulars. First, the beam of light is made to traverse a horizontal layer of air, thereby minimising effects due to change of temperature, and second, that a differential method is used in measuring the displacement of two reflected images, one from the surface of a horizontal mirror fixed to the frame of the mercury bath and the other from the movable surface of mercury. The apparatus, which is capable of recording tremors and changes in the vertical, has a sensibility of $0''\cdot05$.

A simple form of the same device has been used by Mallet, Abbot, and other observers in the determination of the velocities with which earth waves are propagated, and it has been clearly shown that to obtain comparable results from two sets of apparatus the images reflected from the surfaces of mercury must be observed with telescopes or microscopes of equal powers of magnification (see p. 99).

In Japan attempts have been made to record the movements of a mast projecting upwards from a body floating on the surface of mercury or water. The laws

governing small oscillations of a ship may be expressed
by

$$\tau = \pi \sqrt{\frac{\overline{l}}{g}},$$

where $\tau =$ the period in seconds

$$l = \frac{\kappa^2}{GM},$$

κ being the ship's transverse radius of gyration and GM
the metacentric height.

The results, which were seismoscopic in character, were
unsatisfactory.

Levels.—In Japan, for a period of about two years, tri-
daily observations were made upon a pair of astronomical
levels placed at right angles to each other beneath a glass
case on a stone column in a darkened room. Movements
were continually observed, but as these were apparently so
greatly influenced by changes of temperature, and because it
was found, as had been previously observed by M. d'Abbadie,
that when the levels were side by side and parallel their
bubbles might be displaced in opposite directions, the
results were never subjected to a careful analysis.

Although other instruments give more satisfactory
records of earth movements than levels, the remarkable
researches of M. Philippe Plantamour, commenced at
Sécheron, near Geneva, show that levels may at least
be used for the investigation of long period movements of
the ground, and it would certainly be an interesting
experiment to study changes which take place in levels
placed at right angles to an axis of rock-folding in a
district where there is reason to suppose that elevation is
yet in progress.

Pendulums

Under the head of pendulums we come to a large group
of instruments, amongst which we find not only the most
ancient forms of seismometers or seismoscopes, but also

the most modern contrivances in use for recording earthquakes and for detecting minute but slow changes of level. They may be grouped under four heads : ordinary and bifilar pendulums, inverted pendulums, duplex pendulums, and horizontal pendulums.

Ordinary and Bifilar Pendulums

The experiences which investigators in Japan have had with ordinary pendulums have been extensive. Some of the instruments have had lengths of 40 ft. and carried weights of from 80 to 100 lb., while others have been a few inches or a few millimetres in length and carried weights of a few grains. The small instruments were in several instances enclosed in vacuum tubes, in the hope that their motion might reveal the presence of earth tremors. The large pendulums, again, were expected to act as steady points relatively to which the motion of an earthquake might be recorded. By the introduction of suitable frictional resistances, an ordinary pendulum may for small displacements be made relatively dead beat, but it is found that, for nearly all local earthquakes, especially those with a pronounced vertical component of motion, such an apparatus quite fails in giving a steady point.

Now and then when an earthquake commences with a motion which is sharp and decided, a satisfactory record is obtained on a stationary surface. This, however, is of rare occurrence. Much more frequently an earthquake sets the pendulum swinging, and renders the record unintelligible. Also, unless proper precautions are taken, the violent motion of the pendulum may destroy the recording apparatus and anything of a delicate nature that is in its vicinity.

Again, if the pendulum does not sensibly move because the motion of the ground in such cases is so minute, no record is obtained unless it is provided with a multiplying recording index. It was by an attachment of this description, which at the same time controlled the motion of the pendulum, that the late Dr. G. Wagener was enabled to show that movements to which persons are quite sensible were to be measured by quantities often very much less than one or two millimetres.

The author's first arrangement of this type of apparatus, which has had many modifications, is indicated in the accompanying fig. 3.

B is the bob of an ordinary pendulum, here shown as the cross-section of a heavy metal ring. Across the middle of this is a thin metal bar pierced by a hole at p, through which the upper end of a pointer $p\,o\,i$ freely

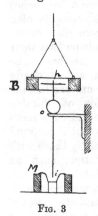

passes. At o this pointer passes through a hole in a fixed arm attached to the box or frame carrying the pendulum. If the pendulum moves, this motion is magnified at i in the ratio of $p\,o : o\,i$, while if the pendulum remains at rest and the fixed arm moves, the magnification is as $p\,o : p\,o\,i$. The motion at i may be recorded by a light needle free to move vertically through guides attached to the pointer, the point of the needle resting on a fixed smoked glass surface (fig. 6). For small earthquakes this apparatus has given satisfactory records. As other pendulums have been devised which give better results, this

Fig. 3

contrivance as a seismograph is no longer used. It is, however, largely used as an instrument to make an electrical contact at the first shake of an earthquake, thereby releasing catches and setting the recording surfaces of seismographs in motion.

In this case the pointer $p\,o\,i$ is terminated by a fine wire leading to o. If i moves in any azimuth it comes into contact with mercury, which stands up as a capillary surface round an iron pin in the vessel M. The mercury and o are connected through a battery with any number of seismographs or recording clocks it may be desired to actuate.

Another use to which the device has been applied is as a tromometer or microseismic indicator. In this case the multiplication of the pointer has been made about 100. The movements of such an apparatus, which is seldom at rest, may be recorded by connections with the drum of a chronograph or by replacing M by a band of paper driven by clockwork, this band of paper being perforated by sparks from an induction coil which, at intervals of about five minutes, are discharged from i. When the long pointer is nearly at rest the diagram consists of a series of holes following each other at regular intervals, but when it is moving the holes form a band with width corresponding to the amplitude of its movements.

An instrument very similar to this was used by M. Bouquet de la Grye in determining disturbances of the vertical at Campbell Island.

In Bertelli's tromometer the stile of a pendulum hanging in a tube is viewed by means of a microscope, and its amplitude of motion is noted as it swings across a scale. Excellent results have been obtained from this class of instrument, but, since different results are obtained from pendulums of different lengths and maxima in storms occurring at irregular intervals, and therefore at any particular instant the general character of the disturbance may not be observed, it fails in giving everything that is desired. In one type of the apparatus a photographic record is obtained by the reflection of a beam of light from a mirror attached to the bob of the pendulum.

On one occasion the writer employed a small silver ball suspended by a silk fibre, the movements of which were noted by the reflection of a ray of light from its surface, which was focussed and then received upon a photographic film.

Messrs. G. H. and H. Darwin, in their researches upon the lunar disturbances of gravity, recorded microseismic disturbances by the movement of a spot of light reflected from the face of a small mirror with a bifilar suspension, one of the suspending fibres being attached to the pendu-

E

lum and the other to a fixed support. The bifilar pendulum
designed by Mr. Horace Darwin, the principle of which
was suggested by Lord Kelvin, has for its bob a circular
mirror about half an inch in diameter. From fig. 4, which
shows the method of suspension, it will be observed that
any tilting of the instrument at right angles to the plane
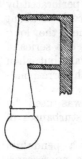
of the paper will result in a rotation of the
mirror and the displacement of a beam of
light reflected from its surface. This portion of the apparatus is enclosed in a tube
of liquid paraffin, the object of which is
to make the pendulum 'dead beat' and
capable of registering earth tilts only.
With the instrument installed at Birming-
ham the sensitiveness is such that it will
indicate a tilt of $\frac{1}{300}$ sec. of an arc, but
it is said not to be impossible to so adjust
the pendulum that a tilt of $\frac{1}{1000}$ of a sec.,
or roughly a slope of 1 inch in 1,000 miles,
might be recorded. An essential feature in the apparatus
employed by Mr. C. V. Boys, F.R.S., in his determination
of the Newtonian constant of gravitation (' Phil. Trans. of
the R. S.,' vol. clxxxvi. pp. 1–72), was a mirror in the
form of a strip suspended from the middle of one of its
longer edges, whilst from its two ends there were hung
fibres of different lengths, each of which carried a small
ball. The fact that a horizontal movement of the point of
support of this arrangement, in a direction at right angles
to the plane of the mirror, causes the latter to rotate,
suggested the idea to Mr. Boys that we have here a new
arrangement which might be introduced into seismometry.

On September 10, 1893, Mr. Boys, whilst observing
with this instrument at Oxford, was disturbed by an
earthquake which had its origin in Turkey.

To return to the pendulums with multiplying indices,
although these are unsatisfactory forms of apparatus with
which to record earthquake motion to which we are
sensible, directly an earthquake has radiated to such a

FIG. 4

distance that the period of the surface undulations to which it gives rise exceed that of the pendulums, the records yielded by an instrument of this type become valuable. It is to the pendulums of Vicentini, Agamennone, Cancani and Ricco, which are from six to twenty-five metres in length and carry weights of 100 to 200 kilos., that we are indebted for many clear open diagrams of movements that have travelled from Japan to Europe. The great objection to this type of apparatus is that it requires for its installation a high building or tower, which is not always obtainable. For the minute tremors which precede the undulatory motion the bob of the pendulum apparently behaves as a steady point for one extremity of the indices, which write their tremulous motion with ink upon a moving band of paper. For the longer period motions the pendulum follows the motion of the supporting building.

With apparatus of this character in which the pendulum is 1·5 metres long, and in which the recording surface is smoked paper, Professor G. Vicentini, in Padua, has obtained a very remarkable series of diagrams of earthquakes the origins of which have been at great distances.

From these remarks it will be seen that although ordinary pendulums often fail to give measurable records of earthquakes near to their origin, they may be advantageously employed as contact makers, and as instruments for recording earthquakes from distant centres, earth-tilting, and the so-called microseismic disturbances constituting tremor storms.

Duplex Pendulums

Although in the early history of seismometry in Japan rows of elastic rods, rising vertically from a board and loaded with different weights at their upper ends, were used in an endeavour to determine by their synchronism, or non-synchronism, with the shaking of an earthquake, the periodicity of its movements, inverted pendulums by themselves, if we except the observations

E 2

made in 1841 at Comrie, have found but little application
in seismometry (see 'Report of the British Association,'
1841). An inverted pendulum so connected with an
ordinary pendulum that the system has feeble stability forms,
however, one of the best types of seismographs which we at
present possess for obtaining static records. From fig. 2
it will be observed that at *o* the recording pointer carries
a small ball, so that relatively to *o* as a centre of percussion
the centre of oscillation of the pointer lies between *o* and *p*.
The result of this is that when the point of support of the

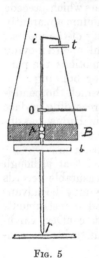

FIG. 5

pendulum together with *o* are suddenly
displaced in any horizontal direction,
say to the left, the bob of the pendulum
is prevented from following its point of
support by a certain restraint exercised
upon it in consequence of the motion
of *o p* being in an opposite direction.

Although the instrument constructed by Dr. G. Wagener must be
regarded as the pioneer among modern
pendulum seismographs, the loading or
o p, rather than *o i*, is a feature in the
construction of these instruments to
which we are indebted to Professor
Thomas Gray, who subsequently drew
attention to the importance of making
an ordinary pendulum more truly
astatic, and suggested methods by
which this might be attained ('Trans.
Seis. Soc.' vol. iii. p. 145).

Professor Ewing's method of attaining this result is
shown in fig. 5. In this arrangement the inverted
pendulum *b* is united with *B* by a ball joint, and the system
so proportioned that, after displacement of the base on
which *b* is resting and a displacement in the same direction of the point of support of *B*, the movements of *B*
tending to the restitution of the system is slightly in
excess of that exercised by *b*; or, if *L* be the length of

the upper pendulum and l the length of the lower pendulum, $B\,l{=}b\,L$ ('Trans. Seis. Soc.' vol. v. p. 89 and vol. vi. p. 19).

The writing index $p\,o\,i$ has a ball joint at p, a universal joint at o connected by an arm with the frame of the instrument, and a hinged writing pointer, the end of which rests on a smoked glass plate, t.

Four forms of instruments embodying the same principles designed and made by myself are shown in the figures :

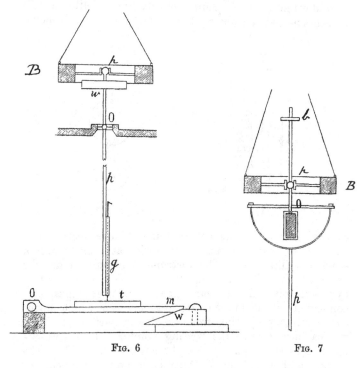

FIG. 6   FIG. 7

If the small ball shown in fig. 3 is increased in size and the arm $p\,o$ in length, then this and fig. 6 become identical. In the latter figure, however, the sliding pointer is shown resting on a smoked plate at i, together with a device consisting of a wedge, w, which may be turned on one side and the shelf carrying the plate dropped sufficiently far so that the latter is no longer in contact with the writing point. In fig. 7 the inverted pendulum b projects above B.

Fig. 8 shows a small form of seismometer which may stand on a mantel-shelf beneath a glass shade six or seven inches in height. In this instrument the maximum motion is indicated by the opening of two scissors-like sets of pointers placed at right angles. One set of these is shown

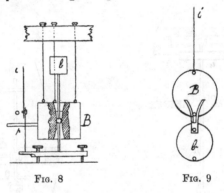

Fig. 8   Fig. 9

centred at o, which are actuated at p by a projecting arm from B, so that the outer extremities at i move over a scale of millimeters. The instrument is only of use for small shocks. The last, fig. 9, is an end view of the two cylinders pivoted on their upper and lower edges, and connected by the branch arm attached to b, which moves over a pivot projecting from the lower part of B. The vertical arm carries the writing pointer. Two arrangements like these, placed at right angles, are used by the author and Mr. John McDonald in an instrument for recording the irregular vibrations and jolts experienced in railway trains.

If we commence with the ordinary free pendulum, then pass to the pendulums with multiplying indices, and finally to those where the indices are used to neutralise the tendency of the pendulum to swing, we have, in the various types of duplex pendulums, a series illustrating the manner in which a useful form of instrument has been gradually developed, and what is true for this type of machine is to a greater or less extent true for many others.

Horizontal Pendulums, Bracket and Conical Pendulum Seismographs

A horizontal pendulum in its simplest form consists of a horizontal bar pivoted at one end on a vertical axis, round which it is free to rotate.

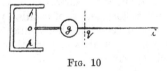

FIG. 10

In fig. 10, if $p\,p$ are the pivots and $o\,i$ the bar—which is usually loaded at some point g, which may be taken to represent the centre of inertia of the system—then, for a small horizontal displacement perpendicular to the plane $p\,i\,p$, rotation will take place round an instantaneous axis passing vertically through some point q when $o\,q = \dfrac{\mathrm{K}^2}{og}$, K being the radius of gyration of the system relatively to $p\,p$. The point q in the arm $o\,i$ is therefore a centre of oscillation, and remains steady relatively to o, which is a centre of percussion and suffers displacement. By pivoting g and making other parts relatively extremely light, we bring g and q into practical coincidence. The displacement of i, which may be a writing point, relatively to o will be in the ratio of $g\,i$ or $q\,i$ to $o\,g$ or $o\,q$. By bringing

g near to o, or by increasing the length $g\,i$, we may evidently obtain a considerable range of multiplication.

After the horizontal displacement of $p\,p$ there is no reason why g should suffer any disturbance, but if such displacement is accompanied by any change in the verticality of $p\,p$ it is evident that there would be a tendency for $o\,g$ to swing round until g occupied its lowest possible position. For the introduction of this principle into seismometry and for its successful application, investigators of earthquake motion have to thank Professor J. A. Ewing, F.R.S.

Because the back and fore movements of an earthquake are not always truly horizontal, it is necessary in order to prevent the wandering which such an arrangement would exhibit, to give the horizontal arm feeble stability. This is done by adjusting the instrument so that the axis $p\,p$ has a slight inclination forwards. A pendulum so adjusted will have a slow period of oscillation.

To get a complete record of horizontal motion which may occur in any azimuth, it is evident that two pieces of apparatus placed at right angles to each other are required.

An important factor in a horizontal pendulum is therefore a steady point or line relatively to which we can measure the motion of points in direct connection with the Earth's surface. Another important function of such an instrument is that it may be employed in measuring changes in the inclination of the surface on which it rests. If these changes are performed rapidly the pendulum required should be short and with no more weight than is sufficient to overcome the friction of its bearings.

Such pendulums will have a comparatively short period. On the other hand, when the changes take place slowly and the angular deviations are small it will be necessary to use a pendulum with a long period.

The sensibility of a pendulum to changes in the vertical will increase with its period, which increases as the axis $p\,p$ is adjusted to approach the vertical.

In 1862 a horizontal pendulum was described by
M. Perrot, and it seems that in 1832 one may have
been used by Lorenz Hengler. As a scientific instrument
it was independently invented in 1869 by Professor
F. Zöllner, who used it in an attempt to measure effects
due to lunar attraction. With this instrument—which
consisted of a glass rod loaded at one end, while the other
end passed through rings at the end of two steel springs,
one running downwards and the other upwards from
projections which were nearly above each other, at the
top and bottom of an iron stand—the inventor considered
that he was able to detect changes of $0''\cdot00035$ and $0''\cdot001$
('British Association Report on the Lunar Disturbance of
Gravity,' 1881).

Shortly before the invention of Professor Zöllner was
made known to the world, the Rev. M. H. Close, of Dublin,
constructed a pendulum in which a rod takes the place
of the mirror in the bifilar pendulum of Mr. Horace
Darwin (fig. 4).

An instrument which, although it does not claim the
sensibility of the pendulum constructed by Zöllner, is
apparently a much more desirable instrument for studying
effects due to slight changes in level, is the horizontal

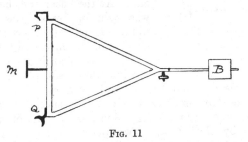

FIG. 11

pendulum of Dr. E. Von Rebeur-Paschwitz (fig. 11). Its
chief advantages are its comparative smallness and freedom
from warping in consequence of inequalities in temperature.
The pendulum, which weighs about 42 grammes and is

58   SEISMOLOGY

about 188 mm. long, is made of brass or aluminium. Its
form is that of a triangle with a small weight at B and two
spherical agate cups at P and Q, which latter rest on short
needle points projecting from a metal stand. At M there
is a small light mirror which reflects light coming through
a slit at a distance of 4½ metres back near to its origin,
where it is focussed by a cylindrical lens to a fine point
and received upon a photographic surface driven by clockwork. When the pendulum is adjusted to have a period
of 18·45 seconds, 1 mm. displacement of the light point
indicates a change of level of $0''\cdot012$.

Horizontal Pendulums used by the Writer

The pendulums used by the writer for indicating or
measuring the unfelt movements of the surface of the
earth have varied in their design according to the nature of

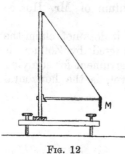

Fig. 12

the motion intended to be indicated. For the purpose of recording earth tremors the pendulum
consisted of 2½ inches of aluminium wire tipped with a needle
point which rested in an agate
cup. At the outer end the boom
was held up by silk or quartz
fibre. Here also there was a
small concave mirror (fig. 12), by
which a vertical beam of light
passing from a slit placed before
a lamp was reflected back to a horizontal slit in a box.
Inside the box, moving slowly, or for special investigations quickly, was a photographic plate or film on a drum
actuated by clockwork. Two pieces of such apparatus
were placed at right angles, with one mirror immediately
above the other, so that they both received the same beam
of light and gave two images about 30 mm. apart on the
film. In one instrument the length of the boom was about
5 mm. and with its mirror weighed only a few grammes.

As instruments for recording tremors they are exceedingly sensitive. In the case of a severe tremor storm the record may be a band 40 mm. in breadth. Since these pendulums cannot be adjusted to have a period much greater than five seconds they are not very sensible to changes in level; nevertheless, it was by the use of such instruments that

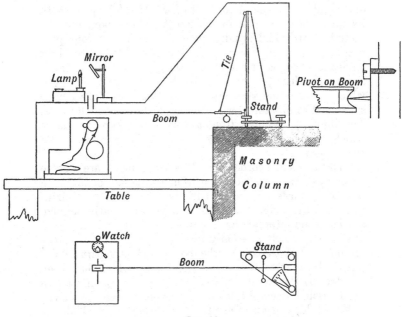

FIG. 13

the writer had his attention first drawn to the fact that there was a diurnal wave in the alluvium of Tokyo.

To study this wave and at the same time to get rid of mirrors, lenses, and all apparatus requiring delicate manipulations, pendulums in which the booms were comparatively long and heavy were constructed. The length of these booms, which have been made of aluminium

tubes or lacquered bamboo, has varied from 2 ft. 6 in. to
5 ft. 6 in. (fig. 13). The one end pivots by means of a
quartz cap resting on a steel point at the base of an iron
stand.   At the outer end is a light plate with a slit
in it parallel to the length of the boom. The supporting
tie for the boom is a thin brass wire. When such a boom
has a period of fifty-five seconds, 1 mm. deflection at its
outer end is equivalent to a tilt of $0''{\cdot}08$.   The plate with
its slit floats over a slit in the top of a closed box, immediately beneath which is a drum covered with a photographic
film driven by clockwork.   As the two slits are at right
angles to each other even when they are at some distance
apart, all that passes into the box is a point of light
which gives a line very much finer than anything I have
succeeded in obtaining with the aid of lenses and mirrors.
The light is obtained from a very small kerosine flame
which, by means of a mirror, is reflected downwards upon
the upper slit.

The tremor records from this instrument are not so
pronounced as those obtained from the small pendulums,
but the records of diurnal tiltings, or those due to earthquakes originating at a great distance, are clearly defined.

The time intervals are obtained from the minute hand
of a watch, which is so arranged that each hour by crossing one end of the slit in the box it eclipses part of the
light.   It is this form of seismometer which has been
selected by the Seismological Investigation Committee of
the British Association for use in a general seismic survey
of the world, and instruments are now established at the
Cape of Good Hope, in New Zealand, North America, and
many other places.

Attempts have been made to photograph a point at
the outer end of a boom, but the resulting diagrams were
wanting in clearness of definition.

An instrument has been designed to record the arrival
of pulsatory motion of large earthquakes and to avoid the
troubles attending photography. It consists of a boom of
four feet in length heavily loaded at its outer end, where

it carries a small pot of ink and a balanced siphon writing
on the surface of paper.   The force of about 1 mg.
applied near to the centre of gravity of such a pendulum
will cause a deflection of about 1 mm.   It has successfully
recorded several earthquakes, but the difficulties in keeping
a uniform flow of ink have been such that the siphon has
been replaced by a light stile writing on the smoked
surface of smooth paper.   All these pendulums, which are
similar in general design to one devised by Mr. A. Gerard,
of Aberdeen, in 1851, are suitable for recording slow
changes in the vertical, diurnal tilting, earth tremors, and
the slow pulsatory motions due to distant earthquakes.
They are also, but to a limited degree, sensible to elastic
tremors.   If these, however, are very slight, like those
produced by traffic in a city, or even by a small local
earthquake, pendulums with long and heavy booms may
remain undisturbed.

Bracket, Conical Pendulums, and Seismographs

The first attempt to employ horizontal pendulums to
record earthquakes was made by Professor W. S. Chaplin,
who, about 1878, employed wooden rods free to turn on
a vertical axis and loaded at their
outer ends.   The arrangement was
without multiplying indices, so that,
in the absence of severe earthquakes,
no records were obtained.   The apparatus was consequently abandoned.
Shortly after this, Professor T. Gray
constructed a pair of hinged horizontal brackets, which at their outer
ends carried extremely heavy weights.

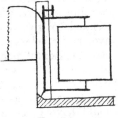

Fig. 14

Following these, came the instru-
ment of Professor J. A. Ewing, F.R.S., in which the bracket
up to a pivoted weight was short, but beyond this point
was continued as a light pointer.   With this instrument
recording on a moving surface, he obtained the first long

open diagram of earthquake motion. The only difference between fig. 14 and Professor Ewing's original design is that the pivots carrying the bracket are so placed that their line of action and that of the weight pass through the same point.

This instrument assumed many various forms, especially in the hands of Professor Thomas Gray, who designed conical pendulums where the upper bracket pivot was replaced by a tie, a bracket seismograph in which the bob is replaced by a ring, which has the advantage of having a considerable moment of inertia, and a double bracket seismograph in which freedom of motion is given to a pivoted mass in all horizontal directions.

All these instruments have been constructed by the writer, and for earthquakes which we can feel and which are unaccompanied by vertical motion they have yielded satisfactory diagrams. When, however, the latter component appears, tilting takes place, and the size of the resulting diagram, which is no longer a representation of horizontal displacements, is practically proportional to the length of the boom, or the length of the boom and its attached pointer. For example, on October 28, 1891, conical pendulums with booms eighteen inches in length which were loaded at their outer ends, where they carried a writing pointer, which were from their construction not intended to give a multiplication of the earth's motion, gave nevertheless a diagram about twice as large as small horizontal pendulums with a bracket and writing index of 9·5 inches in length, or approximately half the length of the long booms. Had each of these types of pendulums acted as steady points, the smaller of the two should have given a record six times greater than that given by the larger.

For earthquakes having this character, and they are not unfrequent, a bracket seismograph, if it did not swing beyond the point which it would reach were it tilted slowly, would become a goniometer, and it was this observation which led the writer to design instruments

capable of recording the undulatory movement of earthquakes (see p. 67).

From these short notes on horizontal pendulums it
will be seen that these instruments, which have been so
useful in modern seismometry, have within a period of
sixty years been independently invented many times
(see the ' Horizontal Pendulum,' by Dr. Charles Davison,
' Natural Science,' vol. viii. p. 233).

Rolling Spheres and Parallel Motion Seismographs

About 1876 Dr. G. F. Verbeck, of Tokyo, constructed
a seismograph in which the steady mass was a heavy
block of wood resting on four crystal balls standing on a
level marble slab. From the upper block a pencil free to
slide vertically rested with its point upon a piece of paper
attached to the lower plate. An arrangement very similar
to this was designed by Mr. C. A. Stevenson, of Edinburgh,
who employed it in recording earthquakes in Scotland,
whilst another one was used by the writer in Tokyo.
Professor Thomas Gray constructed a seismograph with a
multiplying index in which the steady point was the
centre of oscillation of a single solid sphere.

In a second form of apparatus only the segment of
a sphere was employed. From this a vertical arm rose
upwards, on the top of which a heavy mass was pivoted,
the arrangement being such that while the upper pivot
represented the centre of oscillation of the system, neutral
equilibrium was obtained.

In a third form two hollow cylinders with their axes
at right angles and standing on a plane surface were employed. In these the instantaneous axis is for a thin
cylinder near to the highest point of the inner surface of
the cylinder.

Mr. C. D. West drew attention to the fact that astatic
suspension might be obtained by link work, and two
pieces of apparatus were constructed, in each of which a
heavy bar of metal was supported, as shown in fig. 15

('Trans. Seis. Soc.' vol. vi. p. 22). The parallel motion here employed is that known as Watt's. For small displacement it allows the bar to have a straight line motion, which therefore hangs in neutral or feebly stable equilibrium. Other linkages which might be employed, as pointed out by Professor J. A. Ewing ('Trans. Seis. Soc.' vol. vi. p. 25), are those of Peaucellier or Tchebicheff, the latter being preferable on account of the fewness of the joints (fig. 16). In this case the links are replaced by cords, so arranged that the distance between their points of support A A is the same as the distance from the centre of the bar B to the centre of the line A A.

Professor T. Alexander's form of parallel

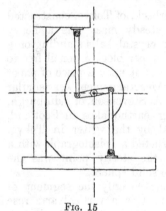

Fig. 16

motion seismograph was as follows. Three small segments of spheres stand in spherical cups of half

Fig. 15

their curvature. Each segment carries a vertical spike, the upper end of which terminates as a diameter to the segment, so that when the segments roll the ends of the spikes move in a plane. A loaded platform which served as a steady plane was carried on the spikes ('Trans. Seis. Soc.' vol. vi. p. 30). The writer obtained a similar result for a mass having freedom in one direction by supporting it on the edges of two comparatively large friction rollers. It will be observed that there is a close connection between the two last types of instrument and the rolling sphere seismographs of Verbeck and Stevenson.

All these forms of apparatus have been constructed

and used, the one giving the best result being that designed by Mr. West. Rolling surfaces present great frictional resistances, and if these are spherical in form there is a tendency to gyrate.

Instruments to Record Vertical Motion

One of the earliest attempts to record vertical motion was made by a committee appointed by the British Association, in 1841, 'for registering shocks of earthquakes in Great Britain.' For this purpose a rod was loaded at one end, and supported at the other by a strong flat spring fixed to a wall. When moved suddenly upwards or downwards the weighted end of the rod by its inertia remained nearly at rest, and was used as the datum from which motion was measured.

Dr. G. Wagener endeavoured to obtain a steady point from a buoy free to rise and fall in a vessel of water.

An improved form of this, experimented upon by Professor Thomas Gray, was a buoy formed like a huge hydrometer, the vertical spindle of which alone projected above the surface of the water. A second form of apparatus consisted of a vessel which had a corrugated or india-rubber bottom, and was filled with water; vertical motion was recorded by using this flexible bottom as a steady point. To obtain a fixed point from a weight suspended by a spiral spring has, on account of the impossibility of obtaining a spring long enough and extensible enough to give a sufficiently long period, been found impracticable. If the weight terminates as a stile and its function is to move and make contact with a surface of mercury close beneath, it becomes a fairly sensitive seismoscope, but beyond this in seismometry this extremely old form of apparatus is without value.

None of the above instruments were satisfactory, and for many years seismologists in Japan were baffled in their attempts to register vertical motion ('Trans. Seis. Soc.' vol. i. p. 48). In 1880, however, the problem was solved

F

by Professor Thomas Gray, who constructed an instrument in which a heavy mass had given to it an exceedingly slow period of vertical motion by fixing it to the end of a lever, the short arm of which was attached to and stretched a spiral or flat spring (fig. 17). Had the spring been stretched by the direct attachment of a weight, its period of vertical oscillation would have been that of a simple pendulum, the length of which was equal to the amount of elongation and may be expressed

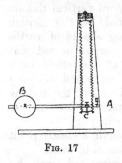

as $\tau = 2\pi\sqrt{\dfrac{E}{g}}$, when E is the normal elongation of the spring. When the elongation is obtained by the horizontal lever the period may be written $\tau = 2\pi\sqrt{\dfrac{El}{gl_1}}$, when l is

FIG. 17

the length of the longer arm of the lever and l_1 that of the shorter arm. Whatever gain there is in the neutrality of the mass is, however, balanced by loss in the effectiveness of its inertia.

To increase the astaticism of the mass carried on the lever, Professor T. Gray attached to the outer end of the bar a hermetically sealed tube containing mercury, which when the bar was depressed ran outwards and increased the load in such a manner as to compensate for its decreased leverage.

Professor J. A. Ewing obtained this desirable compensation by placing the point of attachment of the spring not on the line A B, but a certain distance below it, at C, as shown in fig. 17, by which arrangement, when the spring lengthens or when the moment of B is reduced, the point of attachment moves towards the fulcrum, and *vice versa* when the spring shortens.

The history of the horizontal lever seismographs has been like that of the duplex pendulums, bracket seismographs, and other classes of instruments—an initiative has been followed by a series of modifications and improvements. One of the writer's modifications of Mr. Gray's idea, which he has largely used chiefly in recording motion when great multiplication was not required, has been to use a spring coiled like that in a clock. A spring, which has the form of a spiral, is controlled by the moment of a weight attached by a horizontal lever to a containing box. In this form the instrument becomes extremely small. This type of instrument is at present the best which seismologists have at their disposal, but, like the horizontal pendulums, when it is subjected to tilting in any other direction than that at right angles to the horizontal lever, it is at once affected by gravity and its records are no longer truly seismic.

The author arranged several instruments of this nature in different azimuths, but the records they have given for the same earthquakes have not yet been analysed.

Instruments to Measure Earthquake Tilting

It has already been stated that bracket seismographs at the time of a moderately severe earthquake often suffer displacement in consequence of being tilted, and it seems possible that a boom held in a horizontal position by a tie, and lightly loaded at its end with a writing pointer attached, might be so adjusted that its deflections would fairly well represent the tiltings it might experience.

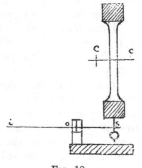

An instrument designed by the writer (fig. 18) for the same purpose consists of a pair of heavy wheels about eight inches in diameter, free to move on knife edges at c, and slightly loaded with a small ball on a stile at s.

Fig. 18

Even for moderately slow tiltings parallel to the plane of one of these wheels s tends to remain at rest.

F 2

Such movements are magnified by the pointers *s o i* pivoted at *o*, which write on the smoked surface of a plate of glass moved by clockwork.

Recording Surface and Writing Pointers·

To record the violent motions of an earthquake or sudden movements such as are experienced on railroads, a pencil point writing on paper may be found sufficient, but if the motion is continuous and extends over a considerable time, it will be found that such a point becomes rounded and glazed, and will require resharpening every five or six hours. With metallic pointers on so-called metallic paper the diagram is faint and glazing takes place more rapidly than it does with pencil. A pointer more suitable for seismographs is one made from aniline pencil, so loaded that it gives a faint mark upon paper, which can be subsequently intensified by dipping the diagram in water.

Writing points of this character fail, on account of their excessive friction, to record small earthquakes and even the preliminary tremors of large disturbances, and are therefore only admissible in instruments where it is not essential to have great sensibility.

The pointer most commonly used is made from a straw, a piece of aluminium, or a fibre of glass. which is so balanced or supported that its outer end rests very lightly on a smoked surface, which may be a disc, a plate of glass, or a drum covered with a sheet of smooth paper or a thin sheet of brass. When the recording surface is a disc of glass driven by clockwork, which is usually set free at the time of an earthquake by electrical connections with a contact maker, there is, when the earthquake continues during several revolutions of the plate, so that the vibrations become superimposed, a certain amount of difficulty experienced in analysing the diagram. Another difficulty lies in the fact that the ends of the indices are describing arcs of circles, and that the diagram is written round the circumference of a circle. Further. than this, if three components of motion are recorded their respective commencements are usually on different radii of the plate, and in all cases, to avoid interference between the different diagrams, the writing points are at different distances from the centre of the plate, and are therefore recording upon surfaces moving with different velocities. These latter difficulties are overcome by a record written on a straight plate, which may be two feet in length, but in this case the record shows vibrations as an open diagram over a period of about one minute only.

Diagrams obtained in this manner are covered with shellac varnish and then photographed by the well known blue process.

The use of a revolving drum presents the same advantages as the straight plate, but in addition the apparatus is more compact and the whole of the motion to which the seismograph is sensible is recorded. The diagram may be fixed, not by pouring varnish over the smoked surface, but upon the back of it. Although the glass disc is largely used in Japan, the drum form of apparatus is the more common.

After an earthquake has taken place all these types of recording surfaces require renewing, and when earthquakes rapidly succeed each other records of all of them may not be obtained.

The pointer with small inertia and with a small amount of friction (see p. 61), and which is therefore the most sensible to slight vibrations, is the siphon recorder, which may be made from a capillary glass fibre. This form of index was used in the Gray-Milne seismograph by Prof. Thomas Gray, and although it has recorded many earthquakes the difficulties it presents in manipulation have led to its abandonment for ordinary work. It is successfully used by Italian workers in connection with long pendulum apparatus, and the author has used it in connection with long horizontal pendulums. In the latter case a gravity siphon made from a fine metallic tube was employed, but great difficulties were experienced in keeping a constant flow of ink. When pointers of this nature are used it is necessary that the paper should be continually moving, but in order to obtain an open diagram, the speed of the receiving surface should, at the time of an earthquake, be increased. If this is not done, an open diagram is obtained at the expense of an enormous quantity of paper. If at any time the paper should cease to move, it usually happens that the siphons continue to flow and the apparatus is flooded by a pool of ink.

The first example of a double speed arrangement in clockwork as arranged for a seismometer is found in the Gray-Milne seismograph. In Italy, Dr. Agamennone and Dr. Mario Baratta, and in Germany Dr. E. von Rebeur-Paschwitz, have each designed apparatus to obtain the same end, the higher speed being controlled by electrical connection.

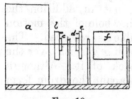

FIG. 19

Although one or two forms have been used in Japan, the following arrangement, suggested to the writer by his colleague, Mr. C. D. West, shown in fig. 19, appears to be the most satisfactory.

A clock a, by means of a pall on b, drives the ratchet wheel c and the shaft connecting it with f, the drum driving a band of paper. On this shaft and turning with it there is a second ratchet d, which may be turned by a pall connecting it with c, which is free on the shaft, but geared to a high speed clock not shown in the sketch.

If, by an electrical contact, this second clock is released, the pall from e acts upon d, and f is turned at a quicker speed than that at which it is usually moving under the influence of a. The last form of recording surface to be described is the photographic film or plate. The writer has worked with plates and Kodak films, but has found rapid bromide paper the material most convenient to manipulate. The various methods which may be followed to project a spot of light or a shadow thrown upon such a surface have already been indicated.

Time Indicators

The ordinary method used throughout Japan to record the time at which an earthquake commences, or at which some particular vibration takes place, is by means of a clock designed by the writer. The three hands of the clock carry three small pads soaked with glycerine ink. The face of the clock is capable of a small to and fro movement given to it by cams on an axle which is turned by a falling weight. This weight is released by means of the contact maker described above (p. 48), which at the same time releases the clockwork driving the surfaces on which the seismographs write their records. When this action takes place by a connection in the same circuit, a small pendulum is set free to swing across a cup of mercury, and at each contact with the mercury a small tick is made upon the edge of the record receiving drum. Many seismographs may be placed in the same circuit with the pendulum, clock and contact maker, and the time of occurrence or period of any vibration may be determined within a very small fraction of a second.

When the records are photographic, each hour or any desired interval of time may be obtained by a short interception either of the source of light coming from the mirror attached to the moving portion of the apparatus, or of a ray of light from a fixed mirror, the beam of which gives a line of reference side by side with that which is recording the earth movements. The method employed by the writer has been described on p. 60.

A Seismograph largely used in Japan

The apparatus shown in the accompanying figure (fig. 20), and designed by the writer, has in its general arrangement features common to the Gray-Milne seismograph. Vertical motion is recorded by means of a horizontal spiral spring seismograph.

Side by side with the pointer for vertical motion are the two light pointers of a pair of bracket seismographs

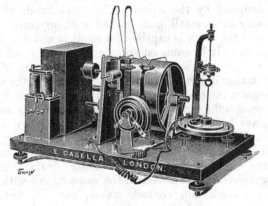

FIG. 20

carried on the same cast-iron stand. The drum on which these three indices write is driven by a spring clock.

The clockwork is released by an electro-magnet freeing the governor, the magnet being in circuit with a clock (not shown), a small battery, and the contact maker on the right hand side of the horizontal pendulum.

The instrument is simple to use, and so long as no tilting is experienced the records are satisfactory.

Microphones

Microphones appear to be still used at certain geo-dynamic observatories in Italy, but experiments made with various forms of these instruments in Japan many years ago showed them to be untrustworthy indicators of small earth movements. Cancani found that as much sound could be heard in the telephone connected with a microphone resting upon a bed of cotton wool, as in one connected with a similar instrument installed upon rock, whilst the writer found very marked difference in the indications of two such sets of apparatus arranged side by side.

Perry's Tromometer

The essential feature in this apparatus, designed by Professor John Perry, F.R.S., is a very light spiral spring made from a flat strip of metal which, when it is slightly stretched or contracted, untwists or twists through a comparatively large angle. One end of the spring is attached to a fixed support, whilst the other end is attached to the end of a heavily loaded balance beam. If the arms of this beam are of equal length, then, like a spoon balanced on the edge of a knife, it tends to remain at rest for tilting parallel to its length, and one end of the spring may therefore be regarded as being attached to a steady point. If the arms are of unequal length, and the spiral attached to the longer and heavier arm, the instrument acts as a recorder for vertical motion. The twist of the spring is shown by means of light pointers, or by a spot of light reflected from a small mirror which moves with the coil of the spring to which it is attached.

CHAPTER V

THE NATURE OF EARTHQUAKE MOTION

The general character of earthquake motion—Unfelt motion may continue over many hours—Earthquakes may be recorded in any portion of the world—The amplitude of motion near an origin and at
great distances from the same— Movements accompanied by destruction—Movements underground—Range of motion as determined
from the width of fissures and cracks—Amplitude as deduced from
maximum acceleration and period—Period of vibrations--The
variability of period and its relation to amplitude—The period of
earth waves—Dimensions of earth waves and sea waves—Direction
of motion—Sekiya's model—Direction of motion in relation to
direction to an origin—Duration of an earthquake—The duration of
vertical motion is short.

ORDINARY earthquakes as felt or recorded consist of a
series of easy to and fro movements of small range, which
continue for twenty or thirty seconds. A typical earthquake, such as is noted at any station in Central Japan
several times per year, commences with a series of minute
vibrations, which can be felt only under favourable conditions. These continue for ten or more seconds, to be
followed by a shock of considerable range and a series of
irregular movements, among which one or two other shocks
may occur, the whole disturbance dying out in slow pulsatory waves which grow longer in period as their amplitude
diminishes (see fig. 21).

The duration of the phenomenon as tested by our sensations may have been from three to six minutes. An
ordinary seismograph, however, shows that there has been
continuous movement for about twice this period. A good
horizontal pendulum without the frictional resistance of a

writing pointer, commences its excursions at about the
same time as an ordinary recording seismograph, but may
continue fitfully to move for one or more hours. The two
extremes of the earthquake diagram are in some respects
analogous to the invisible portions of the solar spectrum,
and have still to a large extent to be investigated.

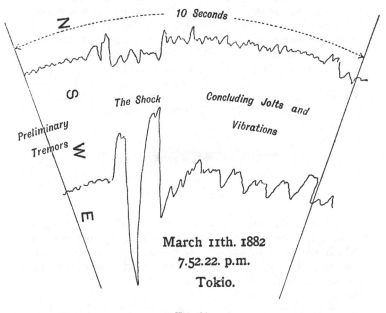

March 11th. 1882
7.52.22. p.m.
Tokio.

Fɪɢ. 21

Underground the vibrations are smaller and much
smoother than those recorded on the surface.

The most terrible form of earthquake is that where the
ground receives a bodily displacement, and is thrown into
undulations, straining the surface even on open plains
beyond the limits of its cohesion, so that fissures are produced.

At a distance of one or two hundred miles from the origin, an observer experiences a long, easy swinging motion, and feels that he is being tilted from side to side. Various observations lead to the conclusion that long flat waves, like water waves, are passing beneath his feet.

Although it has usually happened that with this class of violent earthquakes seismographs installed near a centrum have only succeeded in recording the commencement of one of these disturbances before they were buried in ruins, an excellent diagram of an earthquake closely approaching this type was obtained in Tokyo on June 20, 1894 (p. 138).

A curious form of earthquake is one that cannot be felt. You go to your observatory, and upon each of the recording surfaces of the seismographs you are surprised to see a beautiful series of long period waves.

FIG. 22
[Half-size]

These earthquakes, which seem to be the outer fringes of violent disturbances, are often recorded by instruments which are used to measure slight changes in level. In Japan, with horizontal pendulums, I once recorded ten of these gentle undulatory movements in twenty-four hours.

A photogram of one of these disturbances obtained on June 3, 1893, in a cave at Kamakura on the solid rock, is shown in fig. 22. The actual record is here reduced one-half. It shows at least fourteen points of maximum motion, and has a duration of five hours and twenty-six minutes.

Fig. 23 is the record of an earthquake obtained at Carisbrooke Castle in the Isle of Wight, the origin of which was in North-eastern Japan. In this case preliminary tremors have a duration of thirty-four minutes. In fact, there appears to be a relationship between the distance of an origin and the interval of time by which tremors outrace the decided motion. On quickly moving recording surfaces, such as are employed in Italy, these movements are shown as open diagrams, and we see that movements which were rapid on an epifocal area are, at great distances, replaced by movements of markedly longer period.

The last type of earthquake to be mentioned is one that excites the imagination, and causes persons to wonder what it is that has happened in the ground beneath their

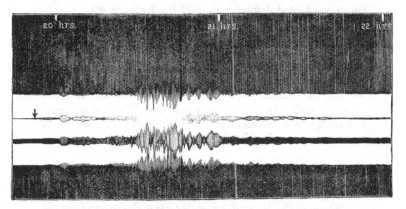

FIG. 23.—JAPAN EARTHQUAKE, CARISBROOKE CASTLE RECORD,
AUGUST 26, 1896

feet. These shocks are so small that so far as I know they have not yet been instrumentally analysed. You are seated, it may be, alone, and all is quiet, when suddenly something like a short, sharp, but gentle blow is felt as if coming from below. There is one little impulse and all is over. In Tokyo these curious little taps seem to be confined to the high ground, but the same impulse has often been simultaneously noticed by persons in different houses. The impression is that a short sharp fracture in the rocks or a small explosion has taken place, but, whatever may be

the origin, the result is that the initial impulse has been
delivered like a blow of a hammer on a small surface, and
this you feel as a quick vibration.

We conclude, then, that although a few earthquakes
are momentary elastic impulses, the greater number of
them consist of vibratory motions propagated within and
upon the surface of the Earth's crust. Partly by reason of
the elasticity of the crust, and partly in consequence of the
influence of gravity, these change in character as they
radiate, the most violent movements near to their origin
being the forced displacements of the rocks or soil.

Later on it will be shown that there are reasons for
believing that the elastic vibrations are due not simply to
compression, but also to distortion.

To render our knowledge about earthquake motion
more definite, we have yet to consider its amplitude and
period, the direction and duration of motion, and the
velocity with which it is propagated. With an accurate
knowledge of these elements we should be in a position to
make a variety of useful and important deductions.

The Amplitude of Earthquake Motion

When at the time of an earthquake the ground moves
to and fro, the diagrams obtained from seismographs
show that a particle has followed an extremely variable
course (fig. 24). The distance between the limits of its
swing is called the range of motion.

The amplitude, which in simple harmonic motion is
half the range, may be defined as the distance between the
position of rest and the limit of the excursion of the particle, whether this be horizontal or vertical.

From what has already been said, it will be inferred
that the range of motion in a given earthquake at the
same station varies from its commencement to its end.
The minute preliminary tremors have a range of motion
measured by a small fraction of a millimetre, while the

pronounced movements may have a horizontal range of
motion anything between a millimetre and a foot.

A movement of 1 or 2 mm. will be strongly felt;
if it reaches 10 mm. it is dangerous; while if it exceeds
20 mm. it is certain to be accompanied by the shattering
of chimneys and by other forms of destruction.

The late Professor Sekiya, after examining diagrams
of 100 shocks recorded in Tokyo, 28 only of which were

FIG. 24.—JAPAN: SEPTEMBER 3, 1887
[Multiplication = 6]

accompanied by a measurable vertical component, found
that the *average* maximum horizontal motion and the
average maximum vertical motion were respectively
1·2 mm. and ·18 mm., while the average ratio between
these two quantities in earthquakes characterised by both
kinds of motion was approximately as 6 : 1.

With a horizontal motion of less than 1 mm. vertical

motion is rarely recorded, and excepting for large earthquakes it is visible only in diagrams which have been obtained near to an origin.

The following are examples of disturbances recorded in Tokyo and Yokohama which were sufficiently intense to unroof a few buildings, shatter many chimneys, dislodge stones from the face of walls, crack brickwork and plaster, and cause other damage.

February 22, 1880. This earthquake caused considerable destruction in Yokohama. Many brick chimneys fell, several houses were unroofed, tiles were dislodged, the main timbers in certain roofs were broken, gravestones were rotated, while bodies like the tops of stone lanterns and the corner-stones of chimneys were projected.

The range of horizontal motion appears to have varied from 15 to 50 mm.

In Tokyo, which is about eighteen miles distant, the movement was 21 mm. on hard high ground.

October 15, 1884. On this day one or two chimneys fell in Tokyo, plaster fell from ceilings, and several brick walls were cracked.

At Stotsbashi (former site of Tokyo University), where the ground is soft, the maximum amplitude recorded was about 13 mm.

January 15, 1887. This disturbance originated about 35 miles S.W. of Tokyo, and near the origin it destroyed many storehouses built of wood and clay and covered with heavy tile roofs, and opened fissures in the ground.

In Yokohama, about ten miles from the origin, the range of horizontal motion was about 35 mm., and a few buildings were destroyed and chimneys shattered.

In Tokyo the ranges of motion recorded were—at Stotsbashi, where the ground is soft and damp, 25 mm.; at the University in Hongo, where the ground is dry and hard, 7·3 mm.; and at the Imperial Observatory, where the conditions are apparently very similar to Hongo, 19·2 mm.

June 20, 1894. On this day Tokyo experienced one of the most severe shakings which have occurred in the vicinity during the last forty years. Nearly all buildings suffered slightly, the commonest form of destruction being the cracking of chimneys or the loosening of tiles, while many substantial brick buildings situated on high ground, like the British and German legations, were so far shattered that it became necessary to rebuild. The greatest destruction occurred amongst foreign built churches and schools situated on the low flat ground of Tsukiji, the foreign concession at the head of Yedo Bay.

One seismograph only, which was without multiplying indices, wrote a diagram. All others had their pointers thrown beyond the recording surface.

The maximum horizontal motion indicated was 63 mm. and the vertical motion 10 mm. (fig. 33).

The ranges of motion given in the above few examples were obtained from the movements recorded upon short thick columns of masonry rising from solid foundations such as are used for the installation of seismographs. It must be borne in mind, however, that they do not necessarily represent the maximum movement experienced throughout Tokyo and Yokohama, for experiments have most definitely shown that stations which are comparatively near to each other give, especially for the stronger shocks, diagrams which are very different, and it is an easy matter to select two stations within 1,000 feet of each other where the average range of horizontal motion at the one station shall be five times, and even ten times, greater than it is at the other.

Usually the greatest swing is experienced on soft ground, or along the upper edge of steep slopes, while on the hard ground and along the foot of such steep slopes it is comparatively small (see chapter on 'Construction,' p. 145).

Again, the movement ten or twenty feet underground is somewhat less than it is upon the surface. Although at the time of the 1891 shock, motion was observed in many

G

mines, the usual experience is that, even with disturbances which are sufficient to shatter chimneys, no movement is observed underground, and if it is perceptible it has only been so in shallow levels. It can hardly be supposed that the transmission of earthquake energy is confined to a layer so shallow that it does not include the depths to which mines are excavated. We are therefore forced to the conclusion that the amplitude of motion underground is too small to produce effects which appeal directly to our senses. In a city where there are artificial or natural differences in soil and contour an earthquake may show in its effects as marked differences as are shown in the effects of a tidal wave rushing through an irregular archipelago of islands. Should it sweep over a large country where there are alluvial planes, rocky mountains, river courses, lakes, and other surface variations, the diverse effects produced by the flood of motion are yet more remarkable.

Although, according to Mallet, the Neapolitan earthquake caused towns upon hills to rock like the masts of ships, my own observations in Japan show that hill tops are comparatively steady.

In 1891 destruction was marked on every plain between Tokyo and Nagoya, a distance of 200 miles, while those who from the dividing ranges looked down upon the clouds of dust and smoke rising from fallen and burning towns, themselves suffered little if any damage.

The extent to which the measurement of amplitudes which have been given may be considered accurate is referred to in the previous chapter.

A point connected with the movement of an earth particle which should not be overlooked is that the movement toward one side of the central position is, especially with a strong shock, often greater than it is towards the other side (fig. 27).

With a disturbance produced by the explosion of a charge of dynamite the direction of greatest motion is usually inwards or towards the origin of the shock.

At the time of the Gifu disturbance in 1891, Prof. F. Ōmori observed that columns like gravestones and stone lanterns on the eastern side of the southern part of the disturbed area fell westwards or towards the coast, while on the western side of the plain they fell in nearly an opposite direction, or towards the east.

These directions of overturning were therefore inwards or towards an origin running down the plain in a north and south direction, the most sudden motion having probably been inwards or towards the origin. A similar rule seems to have been followed on June 20, 1894, when Tokyo was severely shaken.

We have no instrumental records which give the maximum range of motion for large earthquakes, but from the width of fissures in walls of masonry and the width of cracks formed in the soil of open plains where the extent of swing has overcome the cohesion of the material in which it was taking place, it would appear—as, for example, in the disturbance of 1891—that the movement may have reached nine to twelve inches.

Such a disturbance reaching the face of a cliff or river embankment causes fracture parallel to the unsupported surface, and the resulting fissures may be twenty feet or more in width, the widest ones being nearest to the free face. From the face itself material may be projected.

Another method by which a range of motion outside the limits of a seismograph may be estimated depends upon the measurement of acceleration and period.

From the chapter on acceleration it will be seen that we have every reason to believe in the estimate of this quantity as determined from the dimensions of a body that has been overthrown, while the period of motion may be obtained from a seismograph which, before it has been buried beneath a ruin, has succeeded in recording the earlier part of a disturbance. Taking, for example, the period of the Nagoya-Gifu earthquake at 1·5 seconds, as shown upon several diagrams, and the acceleration at Gifu and Nagoya at 3,000 mm. per second per second, it follows that the range of motion was thirteen inches, a quantity practically not different from that determined by the width of fissures on an open plain.

Period of Earthquake Motion

By the period of an earthquake vibration is meant the time it takes to perform a back-and-forth movement or one complete oscillation.

The shortest periods yet noted belong to the preliminary tremors of a disturbance, and to the ripples often superimposed upon the large waves of an earthquake.

These periods appear to vary from $\frac{1}{5}$ to $\frac{1}{25}$ of a second. Still smaller vibrations may have still shorter periods, and be explanatory of the sound phenomena which precede or accompany certain disturbances.

Pronounced movements or shocks, when the range of motion is from 10 to 20 mm., have periods of from 2 to 2·5 seconds, the longer periods being recorded on the softer ground. Movements of 1 or 1·5 mm., which constitute a considerable portion of many shocks, have periods of about 1 second.

The period of a vertical motion of 2 mm. is about half a second. The terminal vibrations, which are in character very much like the long rolling motion of an earthquake after it has travelled a great distance, have periods varying between 1·5 and 4 seconds, the latter evidently being shown by the forced tilting of the seismograph. When the period is more than 4 seconds we feel no vibration, although we may observe the swinging of objects like a chandelier or the pointer of a seismograph (fig. 25).

A disturbance that has travelled a distance of about 6,000 miles, or say from Japan to Europe, shows preliminary tremors which *may have* periods of from 5 to 12 seconds, whilst the decided movements have periods of 20 or 40 seconds.

Like the amplitude of motion, the period varies throughout a given disturbance with the soil on which

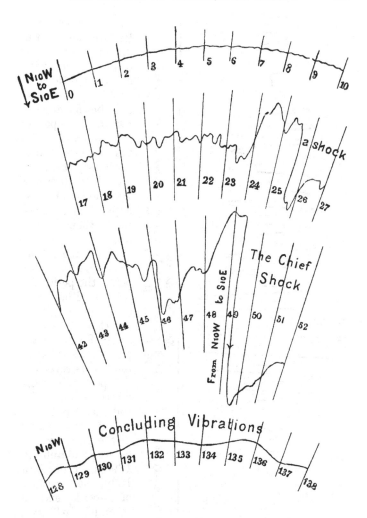

FIG. 25.—MARCH 1, 1882. THE INTERVALS ARE SECONDS OF TIME
[Multiplication = 10]

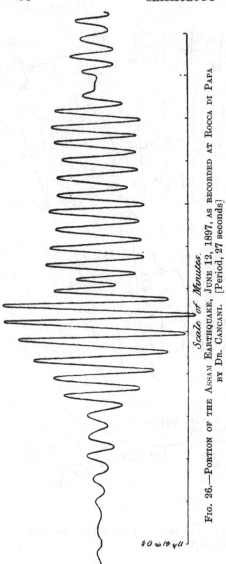

FIG. 26.—PORTION OF THE ASSAM EARTHQUAKE, JUNE 12, 1897, AS RECORDED AT ROCCA DI PAPA BY DR. CANCANI. [Period, 27 seconds]

Scale of Minutes.

the instrument may be installed, the distance from an origin, and with other conditions.

The rules that can be formulated regarding period hold only for the first part of a disturbance, where the record represents vibrations which are approximately elastic in their character.

For such vibrations, up to a certain point, the period increases with, but at a somewhat slower rate than, the amplitude. For the latter part of a disturbance the rule is reversed, the period increasing while the amplitude decreases.

The large waves of the Japan earthquake of March 22, 1894, as recorded in Italy, had periods of about sixteen seconds. In Tokyo, 570 miles distant from the origin, the maximum period was 3·6 seconds.

The Earth Waves of Earthquakes

Beyond the fact that at the time of large earthquakes many persons have seen the surfaces of alluvial plains thrown into a series of undulations, our knowledge about this type of earth movement is extremely limited. The magnitude of these movements has been estimated by the eye, but under such trying circumstances that considerable errors are almost unavoidable.

Mr. Kildoyle, an engineer in Yokohama, who in 1891 was in Akasaka in the midst of the great earthquake, estimates the waves that came rolling down the street in which he was standing as having heights of about one foot, while the distance from crest to crest was from ten to thirty feet.

Observers in Manila describe similar appearances more as ripples a few inches in height and a few inches apart.

At a distance from an origin these earth waves are no longer visible to the eye, but from the observer's feelings and the movements he sees impressed upon various objects, he is compelled to conclude that at intervals of two or three seconds long waves are gently passing beneath his feet.

On the morning of October 28, 1891, although 200 miles distant from the centre of the Gifu earthquake, I was awakened by a powerful rolling motion, which caused pictures to move back and forth, scraping against the wall on which they hung.

In my earthquake room I watched horizontal pendulums heeling quickly first to the right, pausing, and then moving to the left. It was clear that they were not swinging freely, nor were they recording horizontal motion, but by being raised, first on one side, and then on the other, they were measuring the angles through which the supporting column was being tilted.

What I next saw was the motion of the water across a tank eighty feet long, twenty-eight feet wide, and twenty-five feet deep, with nearly perpendicular sides. It rose, first on one side, and then on the other, to a height of four feet.

Phenomena similar to these I have seen on many occasions, and one result of the movement is to produce a feeling akin to sea-sickness.

On March 22, 1894, for one hour and forty-seven minutes, I and my colleague, Mr. C. D. West, watched the fitful and extremely eratic motions of a delicate horizontal pendulum. We could not feel any motion. These movements were not due to the natural swing of the instrument, but were forced displacements, evidently produced by an irregular and intermittent tilting of the bed plate.

The disturbance causing these displacements had its origin on the North-east coast of Yezo, some 570 miles distant, and diagrams of the same earthquakes were obtained in Europe.

Another evidence of tilting is shown in the behaviour of bracket seismographs, which give records proportional to the length of their indices or booms, and not proportional to the multiplication they should show had any part of them acted as steady points (see p. 62).

It is often observed in an earthquake diagram that no two successive waves have the same period.

Dimensions of Earthquake Waves

Our knowledge about the dimensions of earthquake waves is extremely scanty and very imperfect. From what we see and feel we should judge the length of an earthquake wave to be measurable in tens of feet, rather than in hundreds or thousands of feet, to which we are led by calculations from the velocity of propagation.

For example, in the Gifu earthquake of 1891 Mr. Ōmori, in Tokyo, obtained records of small vibrations with a period of one-twentieth of a second. The larger waves had a period of two seconds. The mean velocity of the disturbance between Gifu and Tokyo was about 8,000 feet per second, which leads to the result that there were waves from 400 to 16,000 feet in length.

By calibration of the seismographs at the University laboratory, it seemed that the tilting they experienced at the time of the same earthquake might have been produced by an angular deflection of one-third of a degree. Assuming that this represented the maximum slopes of symmetrically formed harmonic wave surfaces, and that the actual height of such waves was 10 mm., as recorded by seismographs, then the length of the waves which were recorded may have been from eighteen to twenty feet.

We have here two results so hopelessly discrepant that all confidence in such determinations of wave lengths seems to be destroyed.

Waves which have travelled extremely long distances—as, for example, from Japan to Europe—have done so at rates varying between 2 and 10 kms. per second.

The period of these waves as recorded at Rocca di Papa, near Rome, and at Pulkova, is, according to Dr. Adolfo Cancani, twenty-five seconds, from which with a mean rate of 2·5 kms. (8,250 feet) per second would give wave lengths of more than 50 kms. (thirty-one miles).

A sea wave caused by an earthquake travelled 8,778 miles from near Iquique to Japan at a rate of 512 feet per second, and its period near Japan was about twenty minutes. The length of such waves would be about 100 miles. The distance from crest to crest of waves propagated from Japan to San Francisco seems to have been a little over 200 miles.

Without adding to the various estimates which have been made of the length of earthquake waves, a conclusion that agrees with what might be presupposed is that these lengths are extremely different, varying with the intensity of the initial disturbance, the position of the observers, and the nature of the material in which it originates.

Immediately around an epicentrum there are waves of a quasi-elastic character, which may be several inches or even a foot in height, and which succeed each other at distances of a few tens of feet.   At great distances from an origin, the distance from crest to crest of corresponding disturbances, the propagation of which is also influenced by gravity, may be measured by a few tens of miles.

The length of a wave that is nearly truly elastic, the existence of which is evidenced in short period preliminary

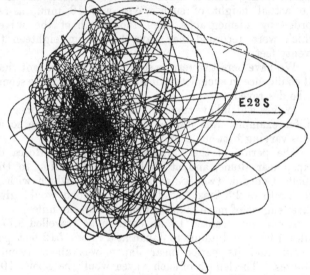

FIG. 27.--MARCH 1, 1882
[Multiplication = 10]

vibrations, must be clearly distinguished from the length of undulatory disturbances of the surface.

Direction of Earthquake Motion

If we place a seismograph with a single index which writes its record on a stationary plate at a distance of about 100 feet from a charge of dynamite or gunpowder buried in the ground, and then explode the charge, we observe that the index of the instrument first moves outwards and then inwards in the direction of the explosion, after which it suddenly commences to move obliquely or transversely to this direction.

The first or normal movements, which are but slightly influenced by transverse waves, are practically along a straight line, but the others, being the resultant of normal and transverse waves, although usually elliptical, may be circular, or like the figure 8, or have forms more complicated (figs. 24, 27, 28 and 29).

In actual earthquakes where there is a definite shock the direction of this may be well defined, but the remaining vibrations may be performed in all azimuths.

FIG. 28.—OCTOBER 25, 1881   FIG. 29.—MARCH 8, 1882
[Multiplication = 7]   [Multiplication = 10]

One result of this is that a seismogram may often be too intricate to analyse.

An excellent illustration of the exceedingly complicated movements performed by an earth particle at the time of a moderately severe earthquake has been prepared by Professor S. Sekiya, in the form of a model in which, by means of a bent wire, the direction of movement is shown second after second for the disturbance of January 15, 1897.

As the distance from the origin of a disturbance increases, the direction of the principal movements becomes more and more indistinct, and finally disappears, all that is left to form a diagram being a complex of ellipses and other figures. Notwithstanding this, if, instead of looking at the direction of the vibrations taken singly, we look at the general effect of all the recorded vibrations superimposed one upon the other, as in fig. 27, we recognise an average direction of movement more or less clearly defined. On the whole the outline of a diagram is usually roughly elliptical, and the major axis of this will be the direction of the greatest amount of motion.

For 1888 and 1889 the direction of the shocks recorded at the Central Observatory in Tokyo were as follows :

N. to S.	N.N.E. S.S.W.	N.E. S.W.	N.N.E. W.S.E.	E. W.	E.S.E. W.W.	S.E. N.W.	S.S.E. N.N.W.	Uncertain
16	4	12	3	45	5	19	4	101

From this it is clear that for Tokyo at least, although in any given earthquake there may be movements in all directions, there is one direction, namely E. and W., in which motion is more pronounced than in others.

Although in some earthquakes a truly vertical motion may occur, causing objects like lamp chimneys to leave their sockets, lamps themselves to leave their stands, and stoppers to leave the necks of bottles, what is usually recorded is the vertical component of a motion, performed in an oblique direction, the range of motion upwards being greater than it was downwards.   Professor S. Sekiya, who examined 119 earthquakes recorded in Tokyo, the distance of whose origins were known, did not find any connection between the direction in which their origins were situated with regard to Tokyo and the direction of the maximum motions of the ground which were recorded at the University Observatory.

In looking over the tables from which this conclusion is drawn, we observe many cases where the two directions under consideration show a remarkable agreement, while in many others they disagree. The disagreement may be due to reflection, refraction, diffusion, interference of normal and transverse motion as a disturbance travels, or it may be dependent upon the general character of the origin itself.

A series of waves originating with the formation of a

long fissure or fault running in a north and south direction, although it may lie to the north of us, could hardly be expected to give a north and south movement.

From the writer's investigations it appears that a quantity of motion arriving from a distance exhibits the greatest movement in a direction at right angles to the dip of the subjacent and neighbouring strata, this being the direction of easiest yielding, and it is for this reason that Tokyo feels more back-and-forth motion in directions lying between north-east and south-east rather than in any other.

Duration of an Earthquake

From what was said respecting the general character of earthquakes it seems that movement is usually felt from thirty seconds to three or four minutes, but, whatever this interval may be, the duration of the disturbance as shown by a seismograph extends at least over double such an interval. The continuous disturbances recorded by free horizontal pendulums have often extended over intervals of one or two hours and have even had durations of fifteen hours, but, as they usually show several maxima of motion, the inference is that they represent the movement due to a succession of shocks, each of which has taken place before the motion due to the one which has preceded it has died away.

Vertical motion, as indicated by a seismograph, appears only at the commencement of a diagram, and it does not continue so long as the horizontal motion.

Movement is perceptible on soft ground for a longer time than it is upon hard ground.

The average duration of 250 earthquakes of moderate intensity recorded by instruments in Tokyo between 1885 and 1891 was 118 seconds. Seven of these, which were strong, were recorded over periods the average of which was six minutes thirteen seconds.

From these few remarks it is evident that we are in

possession of fairly good material respecting the duration of an earthquake at a given station, but to extend our knowledge we desire to know the duration at different stations as a disturbance radiates.

A movement which, as recorded on a seismograph, has lasted six minutes near to an origin may, after it has travelled two or three hundred miles, have lost so much energy that the record on a similar instrument may last only half that interval of time. At a greater distance the duration would be smaller still, until a point would be reached where the record from an ordinary seismograph would be nil. Beyond this limit, however, movement would be recorded by a free horizontal pendulum, and it continually happens that sixteen minutes after the effects of a violent shake have apparently died away in Tokyo and seismographs have ceased to write, the movement has reached Europe, where delicate instruments are just commencing to announce the arrival of an earthquake which may then have a duration of two or three hours.

In some respects the phenomenon may be compared to the manner in which waves are created when a stone is dropped into the middle of a still pond, the disturbance at the centre having ceased, while circle after circle of undulations are rising and falling on the shore. The total duration of the disturbance is from the moment of the first impulse until all motion has ceased, and with an earthquake of moderate intensity this may be many hours.

As might be anticipated, if the acceleration or maximum velocity recorded at a station is great, then the duration of motion at that station is usually long.

The character of the motion produced by explosions of dynamite and other artificial means is epitomised in the next chapter.

CHAPTER VI

VELOCITY OF EARTH WAVES

Introduction—Observations on artificially produced disturbances (experiments of Mallet, Abbot, Fouqué and Lévy, Gray and Milne)—Observations upon earthquakes where the wave paths have been short (Milne and Ōmori); where the wave paths have been long (Newcomb and Dutton, Agamennone, Ricco, Cancani, von Rebeur-Paschwitz, Milne)—The probable nature and velocity of propagation of earthquake motion (the suggestions of Knott, Lord Raleigh, Lord Kelvin)—The paths followed by earthquake motion (hypotheses of Hopkins and Seebach, Schmidt, and a suggestion by the writer)—Conclusions.

THE object of the two following chapters is threefold. The first is to show the character of the information we possess respecting the velocities with which motion is propagated along the surface, and possibly through the interior, of the earth. Secondly, to epitomise the conclusions to which observations apparently point; and finally, while attempting to explain certain of the observed phenomena, to indicate the importance of the results bearing on the physical condition of our globe to which a more systematic continuation and wider extension of present methods of investigation may possibly lead.

Although during the last twenty years much has been done to throw light upon the manner in which earthquake vibrations and movements are radiated from their origins, yet when an investigator finds for the same earthquake velocities varying between 1 and 12 kms. per second he is naturally perplexed.

The results of early observers have often been set

aside on the assumption of their inaccuracy, but no excuse of this description is open to those who have placed before them the records which have been obtained during recent years with the aid of telegraphs and seismographs. In 1847 the gloom which hangs over this branch of earth physics was partly cleared away by Hopkins, who in a report to the British Association called attention to the fact that a distinction should be drawn between *actual* and *apparent* velocities of waves emanating from a subterranean centrum, and that as disturbances passed through mediums of varying elasticity, reflection and refraction might take place. Whether seismologists were assisted when their notice was directed to the condensational and distortional waves in an isotropic medium is somewhat doubtful. Without pausing to consider effects due to the heterogeneity of the Earth's crust, these two kinds of motion were assumed to exist as separate phenomena, and disturbances with a high velocity of propagation were attributed to one, while those with a low velocity were attributed to the other.

One difficulty requiring explanation is the fact that along the same path earth waves originating from a powerful impulse travel at a higher rate than those resulting from an effort of lower intensity. Another common observation is that near to an origin the velocity of propagation is greater than it is between points at a distance. A still more curious phenomenon is that after a disturbance has shown decrease in its speed of transmission it may show acceleration, and this acceleration cannot be with certainty attributed to its having entered a more elastic medium. Finally, we are confronted with the fact that as an earthquake radiates it is preceded by a series of minute tremors, the velocity of propagation of which is certainly very much higher than that of the main disturbance. Before attempting to give an explanation of these and other curious features in earthquake motion, I shall give some account of what has been observed, selecting from the vast amount of material which is at our command illustrations from experiments upon artificially produced disturbances and from records of actual earthquakes, a series of examples in which personal and instrumental errors have probably been small.

ARTIFICIALLY PRODUCED DISTURBANCES

Experiments of Mallet

In the experiments of the late Robert Mallet conducted at Killiney Bay, Dalkey, and Holyhead (British Association Report, 1861), the initial impulse was caused by the explosion of gunpowder. The electrical contact which caused the explosion released a chronograph, which was stopped by an observer directly he saw by means of a microscope magnifying 11·39 times an agitation caused by the resulting waves in a dish of mercury. After corrections for the intervals of time thus noted, the results obtained were in round numbers as follows :

In wet sand	0·251 kilom. per sec.	
In discontinuous granite	0·398	,, ,,
In more solid granite	0·507	,, ,,
In granite at Holyhead (mean) . .	0·371	,, ,,

The charges of powder employed varied between 25 lb. and 12,000 lb., and with but one exception it was clearly shown that the velocity of wave propagation increased with the force of the initial impulse. For example, at Holyhead the relationship between the quantity of explosive and the resulting velocities was as follows :

Powder in lb. . . .	2,100	2,600	3,200	4,400	6,200	12,000
Velocity in kiloms. per sec.	0·335	0·338	0·310	0·344	0·406	0·418

Experiments of General H. L. Abbot

In 1885, when Flood Rock was destroyed by the explosion of 240,397 lb. of *rack-a-rock* and 48,537 lb. of

H

	Distance in miles	Magnifying power of telescopes	Mercury in agitation, sec.	Velocity in metres per sec.
1 Flood Rock ex. 288,934 lb. of explosive	8·33	14	80	1,577 m. Easterly through drift
2 ,,	16·78	14	104	4,086 m. ,,
3 ,,	36·65	18	35?	4,537 m. ,,
4 ,,	48·52	19	54	5,068 m. ,,
5 ,,	144·89	15±	74	3,958 m. ,,
6 ,,	182·68	750	95+	3,335 m. ,,
7 ,,	42·34	{ 31, 16 }	76, 70	1,335 m. ,,, ; { 5,003 m. } Northerly through rock ; 5,243 m. }
8 ,,	174·37	{ 36, 25/45 }	92, 49+	{ 6,243 m. ; 6,210 m. ,, }
9 Hallet's Point ex. 50,500 lb. dynamite	5·13	6	63	{ 1,180 m. } Through clay and drift
10 ,,	8·33	12	72	{ 2,530 m. } Through water and shore of East river
11 ,,	9·33	6	23	{ 1,378 m. } Through clay and drift
12 ,,	12·76	12	19	1,618 m. ,,
13 400 lb. dynamite	1·17	6	8	1,045 m. ,,
14 ,,	1·17	12	18	2,686 m. ,,
15 200 ,,	1·34	6	?	2,051 m. ,
16 ,,	1·34	12	17	2,662 m. ,,
17 ,,	5±	12	—	1,609 m. ,,
18 70 lb. powder	1·40	6	inst.	{ 378 m. } Charge submerged 5 feet
19 ,,	1·34	6	5	{ 1,694 m. } Charge on bottom in
20 ,,	1·34	12	15	{ 2,565 m. } 30 feet of water

dynamite, the most distant observing station was 182·68 miles off. The instant of the explosion was noted at all the points of observation by means of electrical connections and chronographs, while the arrival of the first tremors and their duration were recorded by observers who watched the disturbance of an image reflected from the surface of mercury.

The Hallet's Point observations, where the initial impulse was due to the explosion of 50,000 lb. of dynamite, and others made in connection with subaqueous explosions at the school of submarine mining at Willet's Point, were conducted in a somewhat similar manner. In the table on page 98, which has been drawn up from the scattered writings of General Abbot, the velocities have been reduced to uniform units.

From the above data it is clear, as Abbot shows, that the rate at which a shock is transmitted increases with the intensity of the initial explosion; that when a high magnifying power has been used, tremors in advance of those revealed by a low power have been noticed, with the result that the apparent velocity in the former case is greater than in the latter, and that the velocity of propagation has been higher through rock than through soft material like drift.

A query put forward by General Abbot is whether still higher velocities would have been recorded had telescopes with a greater magnifying power been used? The answer is apparently in the affirmative, and therefore if we wish to compare the observations amongst themselves, not only must we choose those in which the initial impulse has been the same, but where the observers have employed similar instruments. Comparing observations 10 and 12, but not overlooking the fact that No. 10 was largely transmitted through water, and again 16 and 17, we might conclude that as a wave advances its velocity is diminished; but from the first five observations it would seem that there is at the commencement an increase in the initial velocity until it reaches a maximum, after

H 2

which there is a diminution. This increase in the rate of transmission at the outset of a wave from its origin is again seen in experiments 9 and 11. The difference in the velocities recorded for experiments 18 and 19 may be due to the fact that in the case of the shallow torpedo much of the initial energy was expended in throwing a jet of water 330 feet in height into the air. A point well worthy of notice is that the gunpowder waves had a more gradual increase than those observed in shocks produced by dynamite—in other words, the former had a closer relationship to what is so often observed in the records of actual earthquakes than the latter had.

Experiments of MM. Fouqué and Lévy

In the experiments of MM. Fouqué and Lévy the velocity of vibrations on the surface and underground was determined by recording the intervals between the shock, which was usually produced by the explosion of from 4 to 8 kilos. of dynamite, and the displacement of an image produced by a ray of light on a photographic plate moving with uniform motion. The ray of light was reflected from a surface of mercury at the receiving station. The highest velocity was obtained between a point underground and the surface, along a line 383 metres in length, which gave a velocity of 2,526 metres. In this case the shock was due to an explosion of 8 kilos. of dynamite.

The general results obtained were as follows :

—	Veloc. of first trems.	Veloc. of last trems.
	km. km.	km. km.
1. In granite on the surface . .	2·450 to 3·141	·219 to ·108
2. Underground to surface and underground to a greater depth	2·000 to 2·526	1·212 to ·440
3. In *grès permiens* not so compact	1·190	
4. In limestone from surface to underground . . .	·632	
5. In *sable de Fontainbleau* . .	·300	

The velocity evidently increased with an increase in the amount of explosive employed, and it was greatest in the more elastic rocks.

Mallet's determination for granite (507 m,) agrees fairly well with the second maximum in the photographic record (325 m. to 543 m.).

The second set of experiments, considering the nature of the material in which they were obtained and the smallness of the charges employed, give remarkably high results, the velocity for the first maximum exceeding that obtained by firing a larger charge of dynamite in granite on the surface. In a single experiment to determine the velocity between a lower and a higher level underground, the direction of the wave path is unfortunately not very different from that of the stratification, and the velocity is therefore not comparable with those velocities along paths from the upper level nearly transverse to the stratification between it and the surface. If we accept the results of Mallet's experiments, which show that the velocities in these two directions are in round numbers as 1·8 : 1·0, then we may conclude that the velocity between the lower level and an upper level was markedly greater than it was from the latter upwards to the surface.

These experiments show that the velocity between two points on the surface is less than it is between the surface and a point underground. They also indicate that the velocity with which vibrations are transmitted may vary with the depth of the wave path.

Observations of Velocities in Soil, the Wave Paths not exceeding 600 feet

In the author's experiments, which were commenced in conjunction with Prof. Thomas Gray in 1881 and continued at various times during the next four years, the object was not simply to determine the rate of transmission of earth waves, but also to determine their general character. Usually the movements resulting from the fall of a heavy weight, or the explosion of dynamite or gunpowder, were recorded by seismographs. The weights employed varied from 1,710 lb. to 2,000 lb., while the charges of dynamite, which were exploded in holes eight or ten feet in depth, seldom exceeded 2 lb. Although the ground in all cases except one was soft, the resultant vibrations up to distances of about 600 feet were sufficiently large to be recorded as clear diagrams by bracket and other seismographs.

At various stations, usually in a straight line joining them with the focus of the explosion, seismographs were installed, which wrote their movements on the smoked surface of a long plate of glass, the motion of which was controlled by clockwork. One seismograph was placed so that it wrote the movements parallel to the line of installation. These are called normal vibrations. A second seismograph was arranged to record the movements at right angles to such a direction. These are called transverse vibrations. A vertical lever seismograph was occasionally employed to give the vertical motion. A fourth pointer actuated by an electromagnet in connection with a short pendulum swinging across mercury gave a broken line marking small but equal intervals of time.

By the depression of a contact key, the receiving plates at all the stations were set in motion, the pointers of the seismographs drew fine straight lines on the smoked surfaces, while the pendulum indicated intervals of time. A few seconds later a second contact was made and the charge exploded, and the seismographs gave open diagrams of the resulting vibrations. When the earth motion had ceased, all the plates were stopped and were ready to receive a second diagram without any re-adjustments. One observer controlled all the stations, and the only errors due to human interference may have arisen from slight differences in the sensibilities given to the recording instruments. This, however, disappears when velocities were determined, not from the commencement of a disturbance, but from the sharp commencement of individual violent vibrations or from the intervals of time between the appearance of particular waves at the different stations.

Observations were also made with seismographs having single indices, by observing the disturbance created in similar dishes of mercury, and with other arrangements.

Differences in velocity were obtained for each of the three components of motion, and each of these showed a relationship to the intensity of the initial shock, the amplitudes of motion recorded, and the frequency of the waves. It seems advisable therefore, to reproduce the following summary of all the observations. Such conclusions as are indicated do not, of course, necessarily apply to disturbances which have travelled great distances.

I. *Effect of Ground on Vibration*

1. Small hills of alluvium fifty feet in height have but little effect in stopping vibrations.

2. An excavation like a pond ten feet in depth exerts considerable influence in stopping vibrations.

3. In soft damp ground it is easy to produce vibrations of large amplitude and considerable duration.

4. In loose dry ground an explosion of dynamite yields a disturbance of large amplitude but of short duration.

5. In soft rock by the fall of a weight of 2,000 lb. through forty feet it is difficult to produce a disturbance at a distance of twenty feet the amplitude of which is sufficiently great to be recorded on an ordinary seismograph.

II. *General Character of Motion*

1. The pointer of a seismograph with a single index first moves in a normal direction, after which it is suddenly deflected, and the resulting diagram yields a figure partially dependent on the relative phases of the normal and transverse motion. These phases are in turn dependent upon the distance of the seismograph from the origin.

2. A bracket seismograph indicating normal motion at a given station commences its indications before a similar seismograph arranged to record transverse motion.

3. If the diagrams yielded by two such seismographs be compounded, they yield figures containing loops and other irregularities not unlike the figures yielded by the seismograph with the single index.

4. Near to an origin, the first movement is in a straight line outwards from the origin ; subsequently the motion may be elliptical, like a figure 8, or irregular. The general direction of motion is, however, normal.

5. Two points of ground only a few feet apart do not synchronise in their motions.

6. Earthquake motion is probably not strictly a simple harmonic motion.

III. *Normal Motion*

1. Near to an origin the first motion is outwards. At a distance from an origin the first motion may be inwards.

As to whether it will be inwards or outwards is probably partly dependent on the intensity of the initial disturbance, and on the distance of the observing station from the origin.

2. At stations near the origin the motion inwards is greater than the motion outwards. At a distance the inward and outward motion are practically equal.

3. At a station near the origin, the second or third wave is usually the largest, after which the motion dies down very rapidly in its amplitude, the motion inwards decreasing more rapidly than the motion outwards.

4. Roughly speaking, the amplitude of normal motion is inversely as the distance from the origin.

5. At a station near an origin the period of the first wave is shorter than those which succeed it. This difference in period is not so marked at a more distant station.

6. The semi-oscillations inwards are described more rapidly than those outwards.

7. As a disturbance radiates the period of the first wave increases. The period of the second wave seems sometimes to increase and then to decrease. Finally, the period of the normal motion becomes equal to the period of the transverse motion. From this it may be inferred that the greater the initial disturbance the greater the frequency of waves.

8. Certain of the inward motions of ' shock ' have the appearance of having been described in less than no time. This may be explained on the assumption that there was a differential transverse motion between the recording surface and the stand carrying the pendulum recording normal motion (see No. 5 above).

9. Tables have been calculated to show the maximum velocity of normal motion.

10. Diagrams have been drawn to show the ' intensity ' of normal motion.

11. The first outward motion, which on diagrams has the appearance of a quarter-wave, must be regarded as a semi-oscillation.

12. The waves on the diagrams taken at different stations do not correspond.

13. At a station near the origin, a notch in the crest of a wave of shock gradually increases as the disturbance spreads, so that at a second station the wave with a notch has split up into two waves. From this it may be inferred that as a disturbance radiates the number of vibrations which may be recorded may be greater at stations distant from an origin than at stations which are near, and as the period of waves increases with radiation the duration of a disturbance may possibly increase.

14. Near the origin the normal motion has a definite commencement. At a distance the motion commences irregularly, the maximum motion being reached gradually.

IV. *Transverse Motion*

1. Near to an origin the tranverse motion commences definitely but irregularly.

2. Like the normal motion, the first two or three movements are decided, and their amplitude slightly exceeds that of those which follow.

3. The amplitude of transverse motion as the disturbance radiates decreases at a slower rate than that of the normal motion.

4. As a disturbance dies out at any particular station the period increases.

5. As a disturbance radiates the period sometimes increases and sometimes decreases, but this is not marked.

6. As we recede from an origin the commencement of the transverse motion becomes more indefinite.

7. It will be observed that the laws governing the transverse motion are practically identical with those which govern the normal motion, the only difference being that in the case of normal motion they are more clearly pronounced.

V. *Relation of Normal to Transverse Motion*

1. Near to an origin the amplitude of normal motion is much greater than that of the tranverse motion.

2. As the disturbance radiates, the amplitude of the transverse motion decreases at a slower rate than that of the normal motion, so that at a certain distance they may be equal to each other.

3. Near to an origin the period of the transverse motion may be double that of the normal motion; but as the disturbance dies out at any given station, or as it radiates, the periods of these two sets of vibrations approach each other.

VI. *Vertical Motion*

1. In soft ground vertical motion appears to be a free surface wave which outraces the horizontal component of motion.

2. Vertical motion commences with small rapid vibrations, and ends with vibrations which are long and slow.

3. High velocities of transit may be obtained by the observation of this component of motion. It is suggestive of the preliminary tremors of an earthquake.

4. The amplitude and period of vertical waves as observed at the same or different stations have been measured.

VII. *Velocity*

1. The velocity of transit of vertical vibrations near to an origin decreases as a disturbance radiates. Normal vibrations sometimes show a decrease in velocity between the second and third stations, and sometimes show a decided increase. Transverse motions show a marked increase in velocity between the second and third stations.

2. Near to an origin the velocity of transit varies with the intensity of the initial disturbance.

3. In different kinds of grounds, with different intensities of initial disturbance, and with different systems of observation, velocities may vary from 630 (192 m.) to about 200 (61 m.) feet per second.

4. In my experiments the vertical free surface wave has the quickest rate of transit, the normal being next, and the transverse motion being the slowest.

5. The rate at which the normal motion outraces the transverse motion is not constant.

6. As the amplitude and period of the normal motion approach in value to those of the transverse motion, so do the velocities of transit of these motions approach each other.

VIII. *Miscellaneous*

1. At the time of an earth disturbance, currents are produced in telegraph lines.

2. The exceedingly rapid decrease in the intensity of a disturbance, as measured by its power of overturning columns or projecting blocks of wood in the immediate neighbourhood of the epicentrum, has been illustrated by a diagram.

3. For the duration of a disturbance due to a given impulse in different kinds of ground, reference may be made to the detailed descriptions of the first four sets of experiments.

OBSERVATIONS ON EARTHQUAKES

The observations quoted in this section commence with those where the wave paths have not been more than a few hundred feet from station to station. These are followed by the results obtained from instruments separated from each other by distances of from three to six miles, a few hundred miles, and so on, up to velocities determined over paths equal to a quarter of the Earth's circumference.

Observations along Paths of Moderate Length

For several years the author took diagrams of earthquakes at seven stations, each about 900 feet apart. These stations were in electrical connection, so that one pendulum marked time intervals upon each of the moving surfaces upon which diagrams were being drawn. From fifty sets of diagrams, representing fifty different earthquakes, it was only in five instances that the same wave could be identified at the different stations. The result of these identifications led to calculation of velocities of 1,787, 1,302, 1,825, 869, and 501 metres per second.

Even these determinations cannot be accepted without reserve, because it is found that waves spread out as they pass from station to station, a given wave splits up into two waves, &c. Hence a velocity calculated from a wave *a* may be different from that calculated from a wave *b*, which nevertheless is part of the same disturbance. In the diagram from one station a large wave may have a slight notch upon its crest, at another station this notch is seen to have increased in size, while at a third station the single wave appears as two waves. As in the artificially produced disturbances, an earthquake, although it becomes feebler as it radiates, *apparently* increases in its duration.

The same system of observation has recently been elaborated in Japan, but the distances between stations have been increased to several miles. Since the commencement of a disturbance at a given station varies with the sensibility given to the seismograph, the determinations of velocity must be made to depend upon the identification of particular waves upon the diagrams obtained. There must be at least three stations. Up to the present this has only been possible on one or two occasions. On November 30, 1894, at 8.30 P.M., a velocity of 5 kilom. per second was obtained, other disturbances giving from 2·4 to 3·6 kilom. per second.

The following are examples of velocity determinations made in Japan between stations which are in connection with the telegraphic system of the country, and which are provided with seismographs and clocks automatically recording the time and character of particular disturbances. At each of the observatories it is therefore possible to calculate the instant at which a given instrument commenced to write its record.

In 1891, on December 9 and 11, strong shocks originated in the province of Noto on the west coast, which were observed in Gifu, Nagoya, and Tokyo. The mean velocity determined from these records was 2·31 kilom. per second.

The destructive disturbance of October 28, 1891,

which was recorded in Europe, was followed by many after-shocks, the times of arrival of seventeen of which were accurately noted at Osaka, Nagoya, Gifu, and Tokyo. The origin of the main shock was about five miles to the west of Gifu. To reach Tokyo, a distance of 151 miles, the disturbance took 120 seconds. The average time taken for all eighteen shocks was 118 seconds, and the average velocity was 2·40 kilom. per second, the rate of transmission to Osaka being the same as it was over the much longer path to Tokyo.

This same disturbance seems to have reached Shanghai at a rate of about 1·61 kilom. per second, and Berlin at about 2·98 kilom. per second. For the Shonai shock on October 22, 1894, as a mean obtained from observations at ten stations from 60 to 300 miles distant from the origin, a velocity of about 1·95 kilom. per second was obtained.

Giving these last determinations, all of which were computed by Mr. F. Omori, weights proportional to the number of observations each represents, the average rate at which disturbances are propagated over long distances in Japan is 7,560 feet, or 2·3 kilom. per second, a rate which fairly well agrees with that at which the large waves of similar disturbances travel from Japan to Europe.

The conclusion we arrive at from these and similar observations is that from an origin to points a few hundreds of miles distant from the same, movements which can be felt or recorded by ordinary seismographs are propagated at rates of from 2 to 2·5 kilom. per second. To what extent these rates will be increased when the arrival of motion is recorded by instruments capable of recording exceedingly minute tremors has yet to be determined, but it does not appear likely that it will be found to exceed 3 kilom. per second.

Observations along Wave Paths of Great Length

If we collect all the observations which have been made upon the velocities with which earthquake motion has been transmitted to great distances we are confronted with a very large number of contradictory results. For example, Dr. C. Davison shows that for the earthquakes of April 20 and 24, 1894, which originated in North-East Greece and were recorded at forty-one different stations in Europe situated at distances of from 701 to 2,455 kilom. from the epicentral area, the velocities with which motion was propagated to these stations varied between 1·29 and 11·71 kilom. per second. The reason for these discrepancies rests largely on the fact that different instruments, or even similar instruments differently adjusted, have different degrees of sensibility, from which it follows that the phase of motion recorded as the commencement of a disturbance at one station is not the phase of motion recorded as the commencement at another station. If, as in fig. 23, one observer records the commencement of the preliminary tremors, whilst another observer, provided with an instrument less sensitive, only records the maximum phase of motion, there is at once a difference between the observations of thirty-four minutes. Another source of error lies in the difficulty of making accurate determinations of the time at which a movement commenced when the surface on which the record has been made has been moving at a comparatively slow rate. Although we are as yet unable to get a series of records from various parts of the world which are strictly comparable, those contained in the following list, which have been compiled from observations made by Von Rebeur-Paschwitz, the author, the publications of Professor P. Tacchini, and from other sources, may be taken as a series which at least roughly indicates the results towards which more accurate work will lead.

APPARENT VELOCITY OF EARTHQUAKE MOTION ALONG PATHS OF
VARYING LENGTH

Epicentre	Date	Place of Observation	Distance in Degrees	Distance on Arc in Kms.	Velocity in Kms. per Sec. on Arc
1. S. A., Santiago [1] .	Oct. 27, 1894	Tokyo . .	156	17,400	16·0 to 19·0
2. „ „ .	„	Charkof .	119	13,230	12·13
3. „ „ .	„	Rome . .	100	11,200	10·85
4. Merida,Venezuela [2]	Apr. 28, 1894	Charkof .	94·8	10,550	9·1
5. Japan, Sakata .	Oct. 31, 1896	Catania .	88·15	9,796	11·1
6. „ N.E. Coast	June 15, 1895	Ischia . .	87·8	9,749	8·7
7. „ Sakata .	Oct. 31, 1896	Rome . .	86·10	9,564	11·2
8. „ Tokyo .	Nov. 4, 1892	Strassburg .	86·6	9,520	8·1
9. „ Nemuro .	Mar. 22, 1893	Rome . .	86·0	9,500	9·9
10. „ Sakata .	Oct. 31, 1896	Ischia . .	85·3	9,469	11·8
11. „ Nemuro .	Mar. 22, 1894	S. Russia .	„	9,477	8·7
12. „ N.E. Coast	June 15, 1895	Padua . .	84·4	9,320	9·7
13. „ Sakata .	Oct. 31, 1896	Isle of Wight	83·7	9,290	„
14. „ Tokyo .	Apr. 17, 1889	Wilhelmshaven	81·7	9,070	6·8
15. „ „ .	„	Potsdam .	80·6	8,950	11·3
16. Philippines .	Mar. 16, 1892	Nicolaiew .	78·9	8,758	6·08
17. Japan, Tokyo .	May 11, 1892	„	71·2	7,910	9·55
18. „ „ .	Nov. 4, 1892	„	—	—	6·28
19. „ „ .	Jan. 18, 1895	„	—	—	6·3
20. „ Nemuro .	Mar. 21, 1894	Mid Italy .	70·7	7,857	8·2
21. Quetta . .	Dec. 20, 1892	Strassburg .	45·7	5,290	5·65
22. „ . . .	Feb. 13, 1893	„	„	„	3·08
23. Central Asia, Wjernoje	July 11, 1889	Wilhelmshaven Potsdam	43·3	4,806	5·00
24. Quetta . .	Dec. 20, 1892	. —	34·6	3,840	3·86
25. Asia Minor, Amed	Apr. 16, 1896	Strassburg .	18·0	1,990	3·50
26. Patras . .	Aug. 25, 1889	Potsdam .	15·4	1,732	2·59
27. Thebes . .	May 23, 1893	Strassburg .	14·8	1,650	2·4
28. Bucharest . .	Oct. 14, 1892	„	13·0	1,450	2·35
29. Valoria, Epirus .	June 13, 1893	„	12·1	1,350	3·0
30. „ „ .	„	Nicolaiew .	11·4	1,270	3·1
31. Thebes . .	May 23, 1893	„	10·3	1,150	2·0
32. Naples . .	Jan. 25, 1893	Strassburg .	9·0	1,000	3·62
33. Mount Gargano, Italy	Aug. 10, 1893	„	„	„	„
34. Japan, Nemuro [3] .	Mar. 22, 1893	Tokyo . .	8·7	965	2·6
35. „ Noto [4] .	Dec. 9, 1891	„	2·4	272	2·3
36. „ Gifu [5] .	Oct. 28, 1891	„	2·2	241	2·4

[1] Mean of observations at three stations in Tokyo.
[2] Mean of observations at Charkof and Nicolaiew.
[3] Average max. for group of 4 shocks.
[4] Average for a group.   [5] Max. for a group of 18 shocks.

From the following summarisation of the above table
it will be noticed that for distances beyond 2,000 kilom.
the *preliminary* tremors of an earthquake have an
apparent velocity roughly proportional to the distance as
measured on an arc between the origin and the place at
which they have been recorded; in other words, as pointed
out by Mr. J. Larmor, all places at long distances from
an earthquake centrum commence to be shaken at about
the same time.

Distance from Origin		Apparent Velocity in Kilom. per Sec.	
In Degrees	In Kilom.	On Arc	On Chord
20	2,200	2 to 3	2 to 3
50	5,500	5	5
80	8,800	8	7·5
100	11,100	10	8·8
120	13,200	12 ?	10 ?
160	17,700	16 ?	10·5 ?

At present this conclusion can only be taken as a
mnemonic for crude results. The fact, however, that
velocity increases with distances, that for great distances
it is higher than that which we should expect for waves
of compression through a mass of glass or steel, and that
at any observing station only one disturbance is recorded
and not two, which would be the case did waves pass
round our earth, lead to the conclusion that the movements
due to large earthquakes are partly at least propagated
through the world.

Since the duration of the preliminary tremors increases with the distance of the point of observation
from the origin of the disturbance, it cannot be expected
that we should find any great variation in the velocity of
propagation of the heavy motion which succeeds them.
The commencement of heavy motion is sometimes clearly
defined on a seismogram, but it is more usual to find
this phase of motion indefinite, with the result that there
is uncertainty in making exact determinations of the rate

I

at which it has been propagated. Although it has been measured as having velocities of 1 kilom. per second over great distances, it cannot be said to have exceeded 3·5 kilom. per second.

ON THE PROBABLE CHARACTER OF EARTHQUAKE MOVEMENT

If it is assumed that the crust of the Earth has the character of an isotropic elastic solid, then from an earthquake centrum two types of waves may emanate. In one of these the direction of vibration of a particle is parallel to the direction of propagation of the wave or normal to its front as in a sound wave, whilst in the other it is transverse to such a direction, or, so far as this character is concerned, it is like the movements in a ray of light.

These two types of movement, which are respectively known as condensational and distortional waves, are propagated with different velocities, which depend upon certain elastic moduli and the density of the material.

These velocities may be respectively expressed by the quantities $\sqrt{m/\rho}$ and $\sqrt{n/\rho}$ where ρ is the density of the material, n the modulus of rigidity or resistance to distortion, and m a modulus which depends upon the modulus of rigidity and the bulk modulus or resistance to compression k, and is equal to $k + \frac{4}{3}n$.

The first conclusion to which the theory leads is that the condensational wave has a higher velocity than the distortional wave, and therefore ought to outrace it. With artificially produced disturbances *at points near to origins* in fairly homogeneous earth, a phenomenon similar to this has been observed, but whether the preliminary tremors preceding more decided movements observed at great distances represent condensational waves propagated from an origin is yet uncertain. From experiments made by Prof. T. Gray and myself to determine the elastic moduli of granite, marble, tuff, clay-rock, and slate, and the velocities with which normal and transverse movements have been propagated in alluvium, Dr. C. G. Knott drew up the following table as representing average constants involved when determining the velocities with which disturbances may be propagated through fairly solid rocks:

Density . . . . . $\rho = 3$
Rigidity . . . . . $n = 1 \cdot 5 \times 10^{11}$ C.G.S. units
Ratio of the wave moduli . . $m/n = 3$

With the above numbers the velocity of a distortional wave would be $2 \cdot 235$ kilom. per second, while the condensational wave would have a value not quite double this quantity. Should we accept the records made of *decided* movements, which had their origin in Japan and have been recorded in Europe, as representing distortional waves, then our expectations based upon theory closely accord with what has been observed.

On the other hand, small vibrations have been noted which have travelled at rates of from 9 to 12 kilom. per second, a fact which shows that we are not yet in possession of sufficient constants to apply the theory to all the facts which have been observed. Even if we had the constants referring to the elasticity and density of material in the interior of our Earth, we have but to consider the heterogeneity of the materials through which a disturbance probably passes, as Knott and other writers point out, to see that there are serious objections to the assumption that waves with a high velocity are due to the transmission of normal motions while those with a lower velocity represent the less rapid transversal vibrations. At every boundary between two media different in their elasticity, either a condensational or a distortional wave is broken up into reflected and refracted distortional waves as well as reflected and refracted condensational waves, and therefore as a disturbance travels through the heterogeneous mass of materials constituting the earth's crust there is in every probability a continual change in the character of the motion.

Not only does this consideration make it appear unlikely that the tremors which have been observed at stations far removed from an origin, if they were propagated on or near to the *surface* of the Earth, are due to condensational waves, while the more pronounced movements which succeed them represent the distortional waves, but it also indicates that at a given station there should be no definite relationship between the motion of an earth particle and the direction of propagation of an earthquake. For feeble earthquakes, and for those recorded at points outside a meisoseismic area this latter conclusion is remarkably concordant with observation.

On the other hand, however, if preliminary tremors are movements which have been transmitted at *great depths* through a medium where $\sqrt{\varepsilon/\rho}$ is constant or changes gradually, it is likely that they have a condensational character.

Next we may consider the probable nature of surface waves. Lord Raleigh, in a paper on waves propagated along the plane surface of an elastic solid ('Proc. London Math. Society,' vol. xvii. 1885–6), investigates ' the behaviour of waves upon the plane free surface of an infinite homogeneous isotropic elastic solid, their character being such that a disturbance is confined to a superficial region of thickness comparable with the wave length. The case is thus analogous to that of deep water waves, only that the potential energy here depends upon elastic resilience instead of upon gravity.'

Two cases are discussed, but the results are very similar. A particle at the surface moves in an elliptic orbit with its major axis perpendicular. The displacement parallel to the plane surface penetrates into the solid, for an incompressible solid about the eighth of a wave length and to about the fifth into the solid when the Poisson ratio has a value of one-fourth. The surface waves are propagated at a slightly slower rate than a purely distortional wave is propagated.

From observations made upon earthquakes near to

their origin it seems that when vertical motion appears it
is accompanied by horizontal displacements greater than
that required by the formula given by Lord Raleigh, but
the question arises whether the accepted horizontal movements are not more apparent than real—being displacements due to tilting of the recording instruments. That
at the time of a strong earthquake surface waves have an
existence, because they have been seen, been felt, and
been recorded by instruments, is a fact not to be disputed.
As they spread, the distance between crest and crest
apparently increases, and calculations have been made to
determine their height and length. About the path
described by a constituent particle nothing has yet been
experimentally determined. The decided movements which
have been recorded at great distances from their origin,
which have been referred to as possibly being distortional
waves because they slowly tilt pendulums from side to
side, are not unlikely to be *long flat undulations which,
near to the origin, were decided surface waves.* If this is
the case, the phenomenon to be investigated is not the
transversal vibrations of a truly elastic solid, but it is a
quasi-elastic surface disturbance the propagation of which
is accelerated by the influence of gravity.

The preliminary tremors have, however, yet to be
explained. At stations within 100 miles of an origin, as
recorded by seismographs, these outrace the main disturbance, with which, however, they are invariably connected and often overlap it by perhaps ten seconds. At
a distance of 6,000 miles they seem to outrace it by half
an hour.

Knott suggests that they may be due to the quasi-
elastic disturbances which accompany earthquakes. When
the earth movement is violent, and possibly accompanied
by destruction, the material of the earth's surface is
either strained beyond its limit of elasticity, or at least
so far strained that the resulting movements are governed
by co-efficients other than those due to rigidity and com-
pressibility. As these quasi-elastic waves pass through a

region of discontinuity, or as they lose energy, they may be suddenly or gradually transformed into a purely elastic disturbance.

Changes of this description may take place as a disturbance passes from medium to medium, inasmuch as it implies the creation of tremors as the surface waves progress, much in the same way that a trotting pony or a railway train creates the sound waves which run before them, but we seem to be led to the conclusion that the preliminary tremors have a velocity very much higher than those already calculated. This can hardly be accepted, and the only explanation remaining is the assumption that the preliminary tremors are movements originating at an earthquake centrum, and propagated, possibly as condensational waves, along paths yet to be discussed *through* our earth.

Should a more extended and systematic observation confirm this provisional assumption, we shall then be in a position to discuss from a new point of view the physical nature of the materials constituting the interior of our globe which apparently transmits motion at a greater rate than glass or steel.

Points not yet touched upon are the increase of velocity with an increase in the intensity of the initial disturbance, and a decrease in velocity as a disturbance radiates, both of which phenomena are marked near to the origin of artificial disturbances. The only explanation which suggests itself for both these phenomena is that around the epicentrum there is a region to which motion is communicated partly by elastic yielding and partly as a push. The volume of ground which may be thus disturbed is called by my late colleague, Professor T. Alexander, an earthquake core. In the case of an artificial disturbance originating near to the surface, the distance to which this effect extends will depend upon the suddenness and magnitude of the initial disturbance. With an earthquake originating underground, the distance to which a high epifocal velocity may be noticeable will depend not only

upon the above two conditions, but also upon the depth
of the focal origin. The greater this depth becomes,
the greater will be the radius of the epicentral area, in
which there may be not only a real increase in velocity,
but also a high *apparent* velocity.

The conclusion to which these considerations and the
observations which have been made lead is that an earthquake gives rise to at least three kinds of movements,
with different velocities of propagation. On the surface
of the earth there is an undulatory motion, which from
the researches of Lord Raleigh we might expect to travel
at a rate slightly slower than that of a distortional wave;
but, as pointed out by Lord Kelvin, it is probable that
this rate is accelerated by the influence of gravity. What
we should expect and what we find are therefore fairly in
accordance. From a centrum to various points upon the
surface of the earth we should expect truly elastic waves
to be propagated, the velocities of which would vary along
paths of varying depths. At great depths—as, for example,
along a straight or curvilinear path between Japan and
Europe—the velocity of propagation might be higher
than that of a condensational rarefactional wave in glass,
and exceedingly high velocities have apparently been
observed. Lastly, in an epifocal area there may be instantaneous disturbance or an apparent high velocity due to
bodily displacement within an earthquake core and the
transmission of elastic and quasi-elastic vibrations, or to
the combination of such phenomena.

The Paths followed by Earthquake Motion

What has next to be considered are the paths by
which an earthquake originating at a centrum reaches
various points upon the surface of the globe.

Three hypotheses present themselves. Motion may
reach various points on the earth's surface along the
rectilinear wave paths of Hopkins or Seebach, by curvilinear paths as suggested by Dr. A. Schmidt, or, lastly,

by either of these paths; after which, from an epifocal
area the radius of which is not likely to exceed the focal
depth, there is a transmission on the surface of elastic
gravitational waves.

Before discussing the merits of these hypotheses, it
may be well first to consider the case or cases to which
they are applicable.

Since we have no evidence of a disturbance being
simultaneously felt at a number of places on the surface
of our globe and at their antipodes, we may exclude the
idea of a disturbance having originated near to the centre
of our sphere. Other reasons also support this conclusion.
Nor can it be admitted that a disturbance can originate
at half or quarter such a depth, for if it did so, then in an
epicentral area possibly 400 miles in diameter *apparent*
velocities should have been observed which not only would
be enormously high, but would be at least five times greater
than those observed between more distant stations.

From what we know respecting the causes of earthquakes, it is a reasonable supposition to imagine that
their origins are confined to the crumpling of the
material constituting the crust of our globe, which,
according to the Rev. O. Fisher and other investigators, in all probability does not exceed thirty miles
in thickness. The enormous faulting which has accompanied certain disturbances shows that at least a portion
of the initial impulse was delivered actually at the surface.
About the depth to which such faults have descended, or
the mean depth from which an earthquake has originated,
we have no certain information. Confining our considerations to disturbances which have originated at
depths which are extremely small relatively to the radius
of our earth, we may now turn to the hypotheses respecting
earthquake radiation.

In 1847 Hopkins drew attention to the fact that the
velocity with which a wave passes from one point of the
surface of the earth to another point is only an *apparent*
horizontal velocity which may be denoted as v. For

example, if in fig. 30 the origin of a disturbance be O, C be its epicentre on the surface of the earth H' H, and Op_1 Op_2 be the direction of two earthquake rays, then the *apparent* velocity is the distance $p_1 p_2$ divided by the time interval between the observations at the two points $p_1 p_2$. During this interval the distance travelled by the wave within the earth has been sp_2.

The *true* velocity, which may be called V, is that with which it travels within the earth, as, for example, between the centrum and the epicentre. To show the

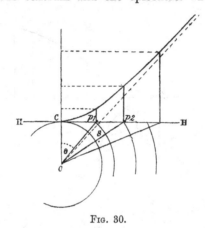

Fɪɢ. 30.

relationship between these two velocities it is assumed that the true velocity is constant. On this assumption, if O is a centrum, wave fronts may be represented by circles or coseismals, the distances between are equal and represent the distance travelled in unit time, which for convenience may be taken at one second. The true velocity is therefore equal to sp_2, while the apparent velocity recorded on the surface is $p_1 p_2$. From the construction $sp_2 = p_1 p_2 \sin \theta$ or $V = v \sin \theta$.

From this it follows that for points near to the epicentre C the apparent velocity is very much greater than the true velocity, while between points at some distance from C the two velocities tend to become equal to each other. The law of this decrease in the apparent velocity is shown geometrically by drawing Seebach's hyperbola, which runs from C through a series of ordinates, the lengths of which are equal to the differences between the time at which C was shaken and $p_1 p_2$, &c., were disturbed. The asymptote to this curve intersects the seismic vertical at the origin, and therefore, if we are satisfied with the hypothesis, having given a number of time observations and knowing

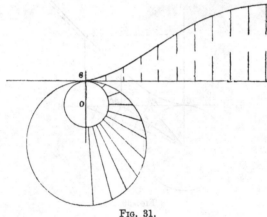

FIG. 31.

the position of the epicentre, the method may be used for determining the depth of a seismic focus.

This hypothesis indicates why a disturbance should apparently be propagated with a high velocity near to its epicentre, but that this rapidly approaches a constant value.

As pointed out by Dr. A. Schmidt, directly we deal with an earthquake which has been propagated over a great distance it is necessary when constructing the velocity curve to take into consideration the curvature of the earth.

This curve (fig. 31), which has lost its hyperbolic
character, shows by the convexity of its upper part that
after a decrease in velocity in the epicentral regions, at
great distances the velocity again increases to become finally
infinite. Dr. Schmidt has likewise shown that actual
observations which have been made upon earthquakes are
best satisfied by a velocity curve drawn on the supposition
that the actual velocity within the earth is not a constant,
but varies with a change in elasticity and density of the
rocks through which the waves are propagated.

If we assume that as we descend in depth there is an
increase in the quantity E/ρ, it then follows that a series

of waves starting from a centrum would be propagated
at a greater rate downwards
than upwards towards the
surface, while the normals to
such a series of waves would
by refraction gradually be
bent upwards.

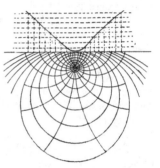

As illustrative of what
would occur under the supposed conditions Dr. Schmidt
gives a diagram like fig. 32,
in which coseismals have been

Fig. 32.

drawn on the assumption that
the velocity has increased proportionately with the depth.
In this case the earthquake rays, which are perpendicular
to successive coseismals, are by refraction turned upwards,
and no longer radiate in straight lines. The coseismals
meet the surface at intervals, which first decrease from the
epicentre and then increase, indicating a decrease and
then an increase in the apparent velocity. The value for
v is never less than the velocity at the centre, but after
rapidly decreasing until it equals this value, it again
increases. The velocity curve or earthquake hodograph,
which shews these changes, is drawn through points determined as they were determined for Seebach's hyperbola.

If there is an increase in the velocity of propagation of earth waves, or in the quantity E/ρ as we descend beneath the surface, whether we take the centrum near to the surface or at a great depth, the resulting hodograph retains its character. The evidence that there is an increase in the velocity of propagation of waves as we trace them beneath the surface is by no means so complete as might be desired (see p. 101). Dr. Schmidt discusses in detail the advantages which the curvilinear propagation presents over that of the rectilinear transmission employed by Seebach.

It will be observed that in fig. 32 there is a great concentration of earthquake rays in the epifocal region, which would correspond to the destruction which is so noticeable in such districts, while with rectilinear radiation the absence of such concentration is not in accord with the results of experience. Although both hypotheses agree in showing a higher apparent velocity near to the epicentre, in Seebach's hyperbola an identical limit is reached for the apparent horizontal velocity for all earthquakes, while Schmidt's modification of the law shows that the apparent velocity on the surface cannot be less than that between the focus and the first coseismal. From this it follows that for limited areas the latter method admits the possibility of very high velocities resulting from earthquakes originating at reasonable depths. With rectilinear propagation, on the contrary, to obtain such high velocities as have been observed it is necessary that origins should be situated at enormous depths.

Should a disturbance originate near the surface, Schmidt's hodograph consists of two symmetrical concave branches which meet in an angle at the centre, indicating that the velocity increases from the epicentre outwards.

The last hypothesis is one that takes into consideration three classes of movements, which immediately round an epicentre are hopelessly confused. These are the truly elastic disturbances which from a focus reach the surface of the earth along rectilinear or curvilinear paths, forced displacements and quasi-elastic waves causing tumultuous movements in the centre of a meisoseismic area, and long undulatory elastic-gravity waves which are propagated over the surface of the earth.

The escape of energy is most pronounced along the paths of least resistance—that is, round the seismic vertical to an epifocal area, and then radially over the surface of the globe. The rate of propagation of the surface waves seems to be about 2 or 3 kilom. per second, and it may be fairly constant. If the minute tremors which have been observed at stations more than 6,000 miles distant from their originating cause, travelled through the superficial crust of the earth, they must have done so at a rate of perhaps 12 kilom. per second, while if they were created as the larger disturbance passed along their velocity, being increased, becomes more abnormal. Reasons have been given for assuming that they came as condensational waves *through* the earth, in which case their velocity becomes 8 or 10 kilom. per second.

REFERENCES

Schmidt, Dr. A. (Stuttgart)	Wellenbewegung und Erdbeben. ' Jahreshefte des Vereins für vaterl. Naturkunde in Württ.' 1888.
„ „	Untersuchungen über zwei neuere Erdbeben, das Schweizerische vom 7. Januar 1889 und das Nord-Amerikanische vom 31. August 1886. ' Jahreshefte des Vereins für vaterl. Naturkunde in Württ.' 1890.
Tacchini, P. . .	Terremoto calabro-messinese del 16 Novembre 1894. ' Reale Accademia dei Lincei,' vol. iii. p. 275.
„ . .	Sulla registrazione a Roma del terremoto calabro-messinese del 16 Novembre 1894. 'Reale Accademia dei Lincei,' vol. iii. p. 365.
Cancani, Dott. A. .	Sulla velocità di propagazione del terremoto di Constantinopoli del 10 Luglio 1894. 'Reale Accademia dei Lincei,' vol. iii. p. 409.
„ „ .	Sugli strumenti più adatti allo studio delle grandi ondulazioni provenienti da centri sismici lontani. ' Reale Accademia del Lincei,' vol. iii. p. 551.
„ „ .	Sulle ondulazioni provenienti da centri sismici lontani. ' Annali dell' Officio Centrale di Meteorologia e Geodinamica,' vol. xv. pt. 1, 1893.

Agamennone, Dott. Giovanni.	Velocità di propagazione delle principali scosse di terremoto di Zante nel recente periodo sismico del 1893. 'Reale Accademia dei Lincei,' vol. ii. p. 392.
,, ,,	Alcune considerazioni sulla velocità di propagazione delle principali scosse di terremoto di Zante nel 1893. 'Reale Accademia dei Lincei,' vol. iii. p. 383.
,, ,,	Alcune considerazioni sui differenti metodi fino ad oggi adoperati nel calcolare la velocità di propagazione del terremoto andaluso del 25 Dicembre 1884. 'Reale Accademia dei Lincei,' vol. iii. p. 303.
,, ,,	Velocità superficiale di propagazione delle onde sismiche, in occasione della grande scossa di terremoto dell' Andalusia del 25 Dicembre 1884. 'Reale Accademia dei Lincei,' vol. iii. p. 317.
,, ,,	Sulla variazione della velocità di propagazione dei terremoti attribuita alle onde transversali e longitudinali. 'Reale Accademia dei Lincei,' vol. iii. p. 401.
Mallet, Robert .	Reports on the Facts of Earthquake Phenomena. 'British Association Reports,' 1851.
,, ,, . .	Report of the Experiments made at Holyhead. &c. 'British Association Reports,' 1861.
Hopkins, William .	Report on the Geological Theories of Elevation and Earthquakes. 'British Association Reports,' 1847.
Newcomb, Prof. Sim. Dutton, Capt., C.E.	The Speed of Propagation of the Charleston Earthquake. 'Am. Journ. of Science,' vol. xxxv., Jan. 1888. *Also other publications lost by fire.*
Fouqué, F., et Lévy, Michel.	Expériences sur la vitesse de propagation des secousses dans les sols divers. L'Académie des Sciences de l'Institut de France. Tome xxx.
Abbot, Genl. H. L. .	On the Velocity of Transmission of Earth Waves. 'Am. Journ. of Science and Arts,' vol. xv., March 1878. *Also other publications lost by fire.*
Milne, J. . . .	Seismic Experiments. 'Trans. Seis. Soc.,' vol. viii.
,, . . .	On a Seismic Survey made in Tokio, 1884–5. 'Trans. Seis. Soc.' vol. x.
Knott, Dr. C.G. .	Earthquakes and Earthquake Sounds as illustrating the General Theory of Vibrations. 'Trans. Seis. Soc.,' vol. xii.
Rebeur-Paschwitz, Dr. E. von.	Horizontalpendel-Beobachtungen, &c., 'Beiträge zur Geophysik,' ii. Band. *Many papers lost by fire.*

CHAPTER VII

SEISMIC ELEMENTS WHICH ARE CALCULABLE

The reliability of the calculations—Maximum velocities and accelerations—Accelerations determined from bodies which have been overturned—West's formula—Lower limits in range of motion to cause overthrow—Ōmori's formula—Experiments in overturning and fracturing—Examples of maximum velocities and accelerations—Mr. F. Ōmori's determinations for the shock of 1891—The necessity to extend the Rossi-Forel scale—Acceleration in a vertical direction—The jumping of stone columns—Intensity of earthquake motion—Acceleration measures 'destructivity'—Isoseismals or lines of equal acceleration—Mendenhall's estimate of earthquake energy—The energy of a cubic mile of earthquake—A practical estimate of the relative energy expended by different earthquakes is the area they shake—The magnitude of an earthquake is connected with the dimensions of its origin.

NEARLY all the facts which have been given in the preceding chapter are the results of direct observation, and as such, under a few possible limitations, they may be used as factors in several calculations which are of practical importance to the working seismologist. In these calculations one assumption is that the back and forth movements experienced at the time of an earthquake are not forced vibrations, but are performed freely with a period dependent upon the constitution of the medium in which they are observed, and may be regarded as having a simple harmonic character.

Since calculations based on such an assumption—for example, those relating to the maximum acceleration as derived from a diagram—agree with the maximum accelerations which have caused the overturning of bodies, and which have been calculated from the dimensions of these bodies, the practical seismologist may place confidence in results which have been subjected to tests of this description.

How far our present knowledge of the nature of earthquake motion would be increased by the careful analysis of vibrations largely magnified and taken on quickly running recording surfaces is still matter for speculation.

Maximum Velocity and Maximum Acceleration

Let $a =$ the amplitude or half semi-vibration in millimetres.

$t =$ the period of a vibration in seconds.

$V =$ the maximum velocity in millimetres.

$f =$ the maximum acceleration in millimetres per second per second.

Then $V = \dfrac{2\pi a}{t}$

and $f = \dfrac{V^2}{a} = \dfrac{4\pi^2 a}{t^2}$

By the maximum velocity is meant the highest velocity with which vibrating particle moves, which in simple harmonic motion occurs half way between its two limits of motion.

This quantity determines the distance to which a body —as, for example, the top of a stone lantern—may be projected.

If we know the height from which a body has fallen and it has been projected, the initial velocity it had to cause it to be thrown the observed distance may be calculated, and this quantity should agree with the quantity $\dfrac{2\pi a}{t}$ as calculated from a diagram.

For artificially produced vibrations such comparisons have been made, and the results show a close agreement.

By the maximum acceleration is meant the greatest

rapidity with which a vibrating particle changes its velocity of motion, which occurs at each limit of its swing.   More popularly, it is the suddenness of the movement in stopping or starting, or, still more briefly, the jerk.

If a body like a column is standing freely on a surface that can be moved horizontally, there is a certain initial rate of movement which will cause the body to overturn in a direction opposite to that of the movement of its base, and this quantity can be calculated from the dimensions of the body.   For this calculation my colleague, Mr. C. D. West, gives the following simple demonstration :

Let the surface of the earth at any instant be undergoing an acceleration of f feet per second per second.   Let M be the mass of a column resting on the ground, y the height of its centre of gravity, and x its horizontal distance from the edge round which it may be supposed to turn.

Then the effect of inertia of the column is as if there existed a force

$$F = Mf$$

acting horizontally through its centre of gravity and tending to overturn the column, the overturning moment being

$$Fy = Mfy$$

This moment is opposed by the moment of the weight of the column Wx, and therefore when the column is on the point of overturning,

$$Wx = Fy = Mfy = \frac{W}{g} f y$$

$$\therefore \frac{x}{y} = \frac{f}{g}$$

$$\therefore f = g \frac{x}{y} \ .$$

If f exceeds this value the column *may* go over, if less the column *may* stand.

K

A less acceleration than $f = g \dfrac{x}{y}$ *may* upset the column if the periodic time of the impulses so far agrees with the oscillation of the column that rocking is established; on the other hand, the same value for f may fail to upset the column if the period is too brief, the impulses then being more shattering than overturning.

If V is the maximum velocity of an earth particle as determined from an earthquake diagram, or by the projection of balls, &c., and t the time of acquiring this velocity, or $t = \dfrac{T}{4}$, where T is the complete periodic time, then the mean-time acceleration is

$$f = \frac{V}{t} = \frac{4V}{T} .$$

If V and t are both very small this formula may be considered nearly correct, but if the amplitude is large then the upsetting value may be nearer to the maximum acceleration $\dfrac{V^2}{a}$ where a is half a semi-oscillation, and not the mean time acceleration.

The results of experimental investigations on this subject are given in a paper on 'Seismic Experiments,' 'Trans. Seis. Soc.,' vol. viii.

The conclusions are that the overturning power of an earthquake, as determined from the dimensions of a body, is at best only approximative. The maximum acceleration of an earth particle apparently lies above the value of f as calculated from the dimensions of a column which has been overturned, and the mean time acceleration lies somewhat below it.

Mallet's formulæ relating to overturning and shattering apparently depend on conditions that do not exist in earthquake movement as recorded in Japan. They are therefore inadmissible.

In the formula for projection, as, for instance,

$$V^2 = \frac{ga^2}{2b} \text{ for horizontal projection,}$$

where a = horizontal distance traversed by the body projected and b = height through which the body has fallen ; the quantity V is apparently identical with the maximum velocity as measured directly or calculated from a diagram, and Mallet's calculations of these particular quantities are of considerable value.

Numerous experiments have shown that the quantity $g \frac{x}{y}$ closely agrees with the maximum acceleration deduced from a diagram of motion in which the motions are such as we meet with in earthquakes.

When the amplitude of motion becomes very small, or the period extremely short, limitations occur in the application of the formula.

According to Mr. F. Ōmori, the following equation gives the lowest limit of the range of motion, when the period is very short, necessary for overturning a rectangular column of height $2y$ and of breadth $2x$, in which $2a$ is the range of displacement :

$$2a = 4x \frac{(x^2 + y^2)}{3y^2}.$$

From which it follows that the range of motion necessary to overturn a column increases with its dimensions. In the experiments on overturning and fracturing, columns of brick and other materials were placed on a truck, which, by means of a connecting rod and a heavy flywheel, could be moved backwards and forwards on its rails through a range and with a period such as might occur in a severe earthquake.

Each back and forth motion was recorded on a band of paper, running with a uniform speed in a direction at right angles to the direction of motion of the truck.

Columns which had to be fractured were clamped to

K 2

the truck at their base. At the instant a column was overturned or fractured a mark was made on the paper, so that the particular wave which was being drawn when this occurred could be identified, and from it the maximum velocity and acceleration experienced at that instant be calculated.

The ratios of the breadth to the height of the columns varied from $1 : 2\frac{1}{2}$ up to $1 : 9$, and in each group there were at least six columns.

These ratios are identical with the dimensional ratios of gravestones and other bodies overturned by the earthquake of 1891.

The actual sizes of the columns were not small, one of them being $9\frac{1}{2}$ inches square and $25\frac{3}{4}$ inches high. The masonry columns, which were built of brick and mortar, or brick with varying qualities of cement, the tensile strength of each of which was tested, were in some cases five feet in height, and usually square in cross-section.

The most important results demonstrated by these experiments showed that columns, whether they are large or small, heavy or light, so long as they have the same ratio of height to breadth, fall simultaneously, and the acceleration recorded as having caused their fall is practically identical with that which may be calculated from their dimensions.

The acceleration causing fracture in a masonry column is given on page 161.

The following are examples of maximum velocities and accelerations, calculated from diagrams of earthquake motion.

For 250 shocks observed in Tokyo between 1885 and 1891, Professor Sekiya found that the average horizontal maximum velocity was 3·3 mm. per second, while the corresponding maximum accelerations were 33·2 mm. per second per second. For seven of the strongest shocks out of the series, these quantities were respectively 22·7 mm. per second and 57·4 mm. per second per second.

The same investigator found that the average maximum

vertical motion and average period for twenty-eight shocks, which occurred between 1885 and 1887, and which were recorded in Tokyo, were respectively ·18 mm. and ·56 seconds, from which quantities it follows that the average maximum velocity and acceleration must have been respectively about 1 mm. per second and 7 mm. per second per second. These quantities are remarkable on account of their smallness.

In Tokyo we have no records given by seismograph where vertical motion has been violent, and yet earthquakes have occurred when lamps have been projected from cup-like stands.

Phenomena of this description indicate an acceleration greater than that due to gravity.

What is probably the most remarkable instance of vertical motion is that referred to by Humboldt, when, at the time of the Riobamba earthquake in February 1797, bodies were projected vertically 100 feet, indicating an initial velocity exceeding 80 feet per second.

It would appear that movements of this description partake somewhat more of the character of efforts which are exhibited when a volcanic vent is established rather than those which accompany earthquakes, and if particles are thrown to a height of 20,000 feet, notwithstanding the resistance of the atmosphere, their initial velocity must have been greater than 1,000 feet per second. Mallet calculated from projective phenomena at the time of the Neapolitan earthquake velocities of 9·1 to 21 feet per second, but in all his calculations it is assumed that the projection took place from rigid supports.

The following table gives similar quantities for shocks which have caused more or less destruction in central Japan.

		Max. vel. mm. per sec.	Max. accel. in mm. per sec. per sec.
Oct. 15, 1884.	Soft ground . . .	68	210
Jan. 15, 1887.	Hard ground . . .	11·5	36
	Soft ground . . .	20	62
	Hard ground . . .	26·2	71·6
Feb. 18, 1889.	. . .	29	83

Since there may be considerable differences in amplitude and period at two places separated from each other by only a few hundred feet, it follows that there will be correspondingly large differences in maximum velocity and maximum acceleration.

The average limits for the former of these quantities, taken at several stations on hard ground and at several stations on soft ground not more than 900 feet distant, were 1·4 and 9·7 mm.

The maximum accelerations experienced, however, varied between 33 and about 100 mm. per second per second.

The meaning of these differences, which are most pronounced with moderately strong disturbances, is that the suddenness of the back and forth motions on one part of the area, where the experiments were carried out, was three times as great as it was upon ground two or three hundred yards distant, from which it may be concluded that buildings upon the area experiencing the most motion might be shattered, while similar buildings not far distant might remain undamaged.

The next illustration is of maximum accelerations which have been determined from the dimensions of bodies overthrown.

These bodies were for the most part gravestones, which exist in hundreds or thousands round every city and village in Japan. They are almost invariably rectangular in section, four or five feet in height, and about two feet square.

These cenotaphs stand on end either freely upon a slab of stone, or else in a slight rectangular recess not more than an inch in depth.

After a severe shaking, such as is experienced in Japan every few years, a temple with its neighbouring cemetery presents a picture of almost indescribable confusion. The temple building, if it has not fallen, is in all probability canted to one side. The tiles are all loosened, while the heavier ones along the eaves have fallen. Stone lanterns, if not lying in fragments round their pedestals, have lost their upper parts, while the lower portions have been displaced and perhaps rotated.

A few gravestones, although they have been displaced and more or less twisted, tell us that the suddenness of motion was not quite sufficient to cause their overthrow, but the majority are piled together with the irregularity of the rocks and boulders met with on many coasts. Each of these has behaved like a column seismometer, and it is no exaggeration to say that after a destructive earthquake in Central Japan there may be a million of these lying on the ground, each of which tells something about the direction of an impulse and its intensity.

After the terrible catastrophe of 1891, Mr. F. Ōmori travelled through the stricken districts, and noted for the many places he visited the general directions in which gravestones and other regular structures had fallen, and from calculations based on their dimensions showed that the shaken country might be mapped into districts in which the average accelerations experienced had been equal.

'In the Neo valley, which was the heart of the disturbed tract, where nearly every building fell, where the ground was faulted and sank vertically or was sheared in a horizontal direction, while forests slipped down from mountain sides to dam the valley and fields were compressed in the ratio of 7 to 10, the accelerations exceeded 4,000 mm. per second per second. In Gifu and its neighbourhood, where accelerations of over 3,000 mm. per second per second were experienced, temples collapsed, and with them 60 to 80 per cent. of all the Japanese buildings fell. The railways were twisted into serpentine forms, bridges supported on cast-iron piers were destroyed, while their foundations were not simply displaced up or down stream, but were brought nearer to each other, the beds of the rivers, like the railroads, having suffered in some places a compression of from 2 to 3 per cent. ; embankments approaching bridges were levelled in the same way that a pile

of sand would be levelled on a shaking plate. Embankments along river courses were fissured as if they had
been cut open by ploughs, making breaches from three to
ten feet in depth. Flat ground was raised into mounds,
and in many places fissures were opened. Earthenware
vessels buried in the ground were broken, while wooden
Japanese bridges, if not destroyed, were so far displaced
that they had to be renewed. Gravestones were overthrown and piled together like heaps of rockery.

'In districts where the acceleration exceeded 2,000 mm.
per second per second a few temples had collapsed, and
although all European brick buildings—which it must be
admitted were not types of good construction—were entirely
destroyed or much shattered, some 10 per cent. of the
Japanese buildings were entirely overthrown, stone walls
were fractured, railway lines were twisted, brick piers
carrying bridges were cracked and had to be rebuilt, while
free surfaces like river banks were fissured.

'Still further away from the origin, where the acceleration appears to have been over 1,500 mm. per second per
second, all Japanese houses, excepting those which were
old and which might have fallen during a typhoon, stood.
European buildings made of brick and mortar suffered
severely, and chimneys of dwelling places and factories
came down. Many stone lanterns and gravestones were
overturned, and river banks and the sides of channels for
irrigation were crushed; but instead of the yawning
fissures, which in severely shaken districts were six or ten
feet in width, in these localities they were not more than
a few inches.

'In localities where the rapidity of change of motion
was about 1,000 mm. per second per second, the destruction
was not so marked in newly built Japanese dwellings;
here and there tiles were disturbed, and old buildings and
weak bridges either fell or were damaged. Brick buildings
and freshly erected chimneys suffered considerably, while
many tombstones and stone lanterns toppled over.

'With accelerations less than 1,000 mm. per second

per second, all well constructed houses and chimneys with-stood the motion. Here and there it could be seen that tiles had been disturbed, especially near to the eaves, while one or two stone columns had fallen over. In localities where the acceleration did not exceed 300 mm. per second per second great alarm had been created. People had sought refuge from their dwellings, which were swaying and cracking, by running outside, where they saw that ponds had become muddy by the lashing of the water.'

Although I had occasion to travel through the shaken district immediately after the great shock, while the secondary shocks were occurring almost every hour and the ruins of fallen towns and villages which had caught fire were yet smouldering, the above notes are based upon the calculations of my colleague, Mr. F. Ōmori, who points out that the limit of the Rossi-Forrel scale of earthquake intensity as used in Europe cannot correspond to anything greater than an acceleration of 2,500 mm. per second per second. For Japan, where the native building stands better than the European structures, to include all degrees of earthquake intensity so far as it is known to us, the scale which is founded on experience in Europe requires extension.

My opinion is that the apparently solid stone structures of an Italian village or the brickwork of a city like London might, with movements having an acceleration of 2,000 mm. per second per second, be reduced to a heap of rubble, while a neighbouring town built of wood, although many of the buildings would be severely strained and a few might fall, would sustain com-paratively little damage.

An exceedingly curious set of observations which Mr. Ōmori made in connection with acceleration in a vertical direction was that in the Neo Valley. Certain objects like gateposts, which were large stone columns, shifted their position by a series of jumps, each leap being from one to four feet. The original position of the gatepost was well marked, while the impressions it made on the ground after each jump were also well marked, and it is evident that the acceleration these objects experienced was something greater than that due to gravity—that is to say, it may have reached 10,000 mm. per second per second.

Not only did stone columns leap, but in one or two instances buildings like Japanese warehouses shifted their position by one or more jumps.

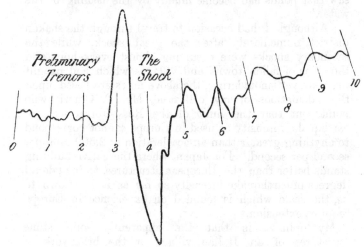

FIG. 33.—THE N.E.-S.W. COMPONENT OF THE SHOCK OF
JUNE 20, 1894
Actual size.   The intervals are seconds of time

Although the story of the accelerations which were recorded of the Nagoya earthquake has here been told, it must not be concluded that our knowledge rests upon this single illustration.

On June 20, 1894, Tokyo experienced a severe shaking, and many buildings and tall chimneys which did not fall were terribly shattered. The destruction, as usual, was particularly noticeable amongst the brick and mortar structures of the Europeans, one noticeable example being the German Legation, which was built with unusual strength. It has now been rebuilt.

In the lower portions of Tokyo, upon the soft ground, accelerations of 1,000 mm. per second per second occurred. A seismograph on hard ground at the University indicated 450 mm. per second per second (see fig. 33).

Intensity of an Earthquake

In popular language an earthquake is usually described as being feeble, strong, violent, or by some term indicating the impression which has been created on the feelings of an observer. Not only is this form of definition inaccurate on account of the varying sensibilities of different individuals, but also on account of the place different observers may occupy at the time of the disturbance. An earthquake which might be strong enough to cause a wooden building to sway violently would to its inmates, especially if upstairs, be considered unusually violent, while if the same persons had been walking in the streets it is quite possible that the same disturbance might have been unnoticed. Again, two persons in buildings of a similar character 100 yards apart might, on account of the difference in the nature of the ground, experience totally different sensations, and what was feeble to one might be strong to another. A consensus of opinions from different districts, and with definitions of what was meant by such terms as ' strong ' and ' feeble,' although by no means giving us all that we desire, would at least convey some general idea about the intensity of an earthquake.

As an example of this classification we may take 3,842 earthquakes which were recorded during the six years between 1885 and 1890 in Japan.

Of these, 2,109 were ' slight,' or only just perceptible ; 1,454 were ' weak,' by which is meant that they were distinctly felt, but were not sufficiently strong to alarm people and cause them to run from their houses.

The remaining 49 were strong, by which is meant that many people sought refuge outside their houses, liquids were thrown out of vessels, certain objects were overturned, while here and there the ground was cracked.

From what we know from determinations based on the records obtained from instruments, the maximum accelerations represented by these three types of earthquakes may have been from 20 to 30, from 40 to 60, and from 200 to 300 mm. per second per second.

It may here be pointed out that these maximum accelerations only measure the ' destructivity ' of an earthquake at a particular station, and they do not represent, nor are they proportional to, the total energy developed at an earthquake origin. If we take, for example, the great earthquake of Lisbon, that of Mino and Owari in 1891, and the Ischian earthquake of 1883, the destruction which occurred near to the origin of these shocks was practically the same, while the radii of their meisoseismic areas were roughly 1,000, 200, and about 10 miles. To compare these three shocks we require to know, not the destructivity at the epicentrum or at a given point in the shaken areas, but the average maximum accelerations at a number of stations throughout the shaken area. With such data destructivity curves may be constructed, the areas of which, between their asymptotes, may be taken as proportional to the relative intensities of the initial impulses.

Absolute values for such intensities might be derived by using the area of the acceleration curve produced by the explosion of a given quantity of dynamite, or the falling of a known weight as a unit.

Measurements of this order are obtainable from earthquake maps, on which the isoseismals are lines of equal acceleration. From a map of this description drawn by Mr. F. Ōmori for the Mino-Owari earthquake it seems that, calling the accelerations experienced in the Neo Valley near the origin 4,000 mm. per second per second, then, at distances of 20, 80, 150, and 300 miles from the origin, the accelerations experienced were respectively 2,000, 1,000, 300, and, say, 200 mm. per second per second.

Although a destructivity curve may be drawn from these numbers, all that it represents is the total motion communicated to bodies on the surface of the earth, and not the energy communicated to the ground. Professor T. C. Mendenhall, in a communication to the American Association for the Advancement of Science in 1888, approaches the question of earthquake energy as follows :

The destructivity as determined at any particular station may be written

1. Maximum velocity $V = \dfrac{2\pi a}{t}$

2. Maximum acceleration $\dfrac{V^2}{a} = \dfrac{4\pi^2 a}{t^2}$

To these he adds :

3. Energy of unit volume with velocity $V = \tfrac{1}{2} d V^2 = \dfrac{2\pi^2 a^2 d}{t^2}$,

(d is the mass per unit volume).

4. Energy of wave per unit area of wave front $= \dfrac{2\pi^2 a d v}{t}$,

where v is the velocity of wave transmission.

5. Energy per second across unit area of plane parallel to wave front (rate of transmission) $\dfrac{2\pi^2 a^2 d v}{t^2}$.

From 5, if A be area of a portion of wave front and l a length measured at right angles to A, then the energy required to generate the waves existing at any moment in the volume $l A$ will be

$$\dfrac{2\pi^2 a^2 d v}{t^2},\ \dfrac{Al}{v} = \dfrac{2\pi^2 a^2}{t^2} \cdot m\ (m = \text{mass of vol. } lA) = \tfrac{1}{2} m V^2 ;$$

' that is to say, the work consumed in generating waves of harmonic type is the same as would be required to give the maximum velocity to the whole mass through which the waves extended.'

In the application of this formula it has to be assumed, first, that the amplitude and period of the subterranean wave does not differ greatly from the same elements of motion observed upon the surface, and, secondly, that the volume of earth which is in motion at any particular instant of time is known.

The following are examples of calculations based on such assumptions :

For the Charleston earthquake of August 31, 1886, assuming a displacement of one inch, a period of two seconds, and a velocity of three miles per second, then the energy of a cubic mile of that earthquake near the epicentrum would be 24×10^9 foot-pounds. To disturb an area 100 miles square the energy would be 24×10^{13} foot-pounds, and the rate of its expenditure would be that of 13×10^{11} horse-power.

In some of my early experiments on an artificially produced disturbance a ball weighing 1710 lb. was dropped thirty-five feet. This represents an expenditure of 60,000 foot-pounds. At fifty feet distance from the place where the ball fell the amplitude of motion was 0·7 mm. and the period about one second. At 150 feet the vibrations were almost imperceptible. The ground was mud and weighed about 110 lb. per cubic foot.

Mr. Ōmori, with these data, shows that the energy per cubic foot was ·00574 foot-pounds, and if a hemispherical volume of radius 150 feet was in motion at any one instant the amount of energy represented would be 41,000 foot-pounds.

When we remember that the movement of an earth particle which on the surface is unconstrained in its vertical excursions is, in all probability, very different from that of a particle deep beneath the surface, that as an earthquake spreads its energy is dissipated in overcoming frictional resistance, and that we do not know the volume of earth that is in motion at the same instant, we must conclude, with the author of this method of analysis, that until more reliable data have been furnished the results obtained by it can only be crude approximations.

Although an absolute measure of earthquake intensity may be unattainable, it by no means follows that we cannot make a rough approximation of the relative amounts of energy expended in different earthquakes.

In a country like Japan an easily obtainable measure of the relative intensities of different earthquakes would be to consider them as proportional to the areas which they sensibly disturbed or which are bounded by similar isoseists.

Other estimates of the relative intensity represented by two different earthquakes observed at the same distance from their origin would be to consider them as proportional either to the maximum accelerations, or to the duration of the disturbances as recorded at the points of observation.

It is not here intended to convey the idea that there is any approximately rigid connection between the area disturbed and accelerations or durations taken as described. These are simply quantities which sometimes have a rough proportionality with each other increasing with the initial effort at a centrum. The quantities which are probably most nearly proportional to the energies developed at different origins are the areas which are shaken.

In connection with estimates of earthquake intensity, observations seem to indicate that the magnitude of an earthquake is directly connected with the magnitude of the fault, or of the material moved beneath the crust by which it is created.

In the case of a large earthquake a large area is suddenly released from a state of strain, resulting in a spring-like motion along a line of fault, followed by an impulse due to the falling or sliding downwards of disjointed strata.

There are no reasons for considering that the time occupied in these operations are for various earthquakes sensibly different, although the time occupied in subsequent settlement may vary within wide limits. For artificially produced disturbances—as, for example, those due to explosion of charges of dynamite in boreholes—observation shows that the rapidity with which surface intensity decreases around the focus is greater for large impulses than for small. A similar law may hold with actual earthquakes, but we have no observations to show that it is marked. The inference, therefore, is that the magnitude of an earthquake effect, unlike that due to an explosion, is largely dependent upon the size of the origin.

CHAPTER VIII

EARTHQUAKES AND CONSTRUCTION

Sites: Soft low ground dangerous—Destruction on high ground rare—Experiences in Ischia—Seismic surveys—Effects on slopes, edges of cliffs, faces of cuttings—Fissures on river banks—Railway embankments. *Foundations*: Regulations in Ischia and Manila—Bridge piers—Movement in pits—Buildings in Tokyo with open areas—Free foundations—Buildings on layers of shot—Lighthouses and aseismatic tables of Stevenson—Van der Heyden's glass house——Japanese houses. *Roofs*: Effect of heavy roofs—Sliding roof in Tokyo—Effect of corbel work—Stability of temples—Support of roofs their pitch—Covering materials. *Walls, Chimneys, Piers*: Italian regulations respecting dimensions—Use of buttresses—End walls—Acceleration which can be resisted by a given wall—Calculation of dimensions—Cast iron and masonry piers—Parabolic piers—Construction of chimneys—Rotation of columns.

It is hoped that the following chapters, which are based upon building regulations and experiences collected from nearly every earthquake-shaken country in the world, the study of thousands of ruined and shattered buildings in Japan, and special experimental investigations respecting the movements required to overturn or fracture constructions of various descriptions, may prove of practical value to engineers and builders whose work is in earthquake districts.

Although the object of each suggestion that is made is to mitigate or avoid earthquake effects, I also endeavour to show what earthquake effects have been and where they have been most marked. To a great extent, therefore, much of what is said may also prove of interest to the student of pure seismology.

Choice of a site.—If in a country like Japan the choice of a site for a city, a reservoir, a building, or

L

other construction was unrestricted, it is certain that
positions could be chosen as free from earthquakes as
many parts of England. Such freedom of choice is,
however, usually limited to a certain area—for example, to
a city. From what has happened to buildings in Tokyo
and from seismometric observations there, we know that,
excepting for local earthquakes, the high hard ground
suffers very much less disturbance than the soft low
ground, so that the city may be divided into two parts,
one of which is comparatively much safer than the other.
The occasions when communities have had their attention
directed to facts of this nature are very numerous, as, for
example, at Lisbon in 1755, Port Royal in 1692, Belluno
in 1873, in Calabria in 1783, San Francisco in 1868,
Talcahuana in 1835, and in Messina in 1726.

Mallet, after his survey of the district devastated by
the Neapolitan earthquake of 1857, states that more places
were destroyed upon the rock than upon loose clay or
other materials, but this, he remarks, may have been due
to the fact that there were more places situated upon the
rock and hills than upon the alluvium and the plains.
Nevertheless he is of opinion that high and lofty situated
places, all other things being equal, are likely to suffer
most. In places situated in an epicentral area, throughout
which a succussatory movement has been experienced, as
was the case in Tokyo on June 22, 1894, it would seem
that movements of the high and low ground, as exhibited
by shattered buildings, have sometimes been practically
equal.

Instances of this description are, however, comparatively
rare, and it is generally found, even for local disturbances,
that buildings on the lower ground have suffered most.

This was so marked after the disturbance of 1883,
which was confined to the small island of Ischia, that the
Government took advantage of the observations which
were then made to mark out sites on which the new town
might be built.

Another point not to be overlooked is the fact that

carefully made seismic surveys have distinctly shown that on a plot of ground not more than ten acres in extent the quantity of motion experienced on one side of such a tract might be sufficient to shatter a building, while a similar building not more than 900 feet distant on the other side might remain undamaged. The dangerous side of one plot of ground, on which, by means of seismographs placed at different points, earth motion was repeatedly measured, bordered a marsh, and was consequently somewhat wet and soft. Marshy, wet ground, which is popularly supposed to absorb earthquake motion, is notably a bad foundation. Experiments show that although the period of motion is lengthened on such ground, the advantage thus gained is more than counterbalanced by the enormous increase in amplitude.

This has been so thoroughly recognised that in the building regulations for Manila respecting structures to resist earthquakes, special reference is made to the character of the foundations which may be used in such places.

In Ischia these rules are even more stringent, there being certain areas of loose soil on which the erection of dwelling places is prohibited. That steeply sloping ground is a bad situation for construction of any description is evident to anyone who has witnessed the effects produced by an earthquake on the faces of steep mountains.

In 1891, throughout the meisoseismal area of Central Japan, landslips were general, the valleys were dammed up to form lakes, while mountain ranges which were green with forest were suddenly left white and naked. Beyond this area, and extending to a distance between 100 and 200 miles from the origin, the effects upon strata resting loosely upon inclined surfaces, although not so severe, were sufficiently well marked to indicate that all steep slopes covered with alluvium were dangerous situations. Not only may material be dislodged from the sides of mountains, but considerable movements are sometimes noticeable on the faces of slopes which have a moderate height.

L 2

At one place along the banks of a river near Nagoya the writer saw a grove of bamboos and other trees which had been moved sixty feet, and the bamboos and trees were yet upright. Along the base of steep slopes there is danger of burial by the falling and sliding of materials from above.

Danger is also to be apprehended along the upper edges of cliffs, scarps, and natural or artificial open cuttings. On river banks the materials adjacent to the free face, being unsupported on that side, swing forwards beyond the limits of their cohesion and separate from the material behind. The general result of a series of repeated movements is not simply to dislodge materials from the face of the cliff or scarp, but also to form a series of fissures parallel to its length, the theoretical distance between which is the amplitude of the wave of motion. This action accompanies all great earthquakes, and is even marked with earthquakes whose range of motion on continuous ground has not exceeded two inches.

In the Aichi prefecture, which stretches eastward from the origin of the disturbance of 1891 more than fifty miles, and is from twenty to thirty-seven miles broad, more than 400 miles of river banks, water trenches, and roads were destroyed by action of this description.

At many places the fissuring was so great parallel to free faces of river banks that they had the appearance of having been destroyed by gigantic ploughs, which had torn out furrows several feet in width, and from a few inches to twenty feet in depth. The distance back from a river front to which these trench-like cracks and hummocked ground extended was from ten to fifty yards. Buildings and roads along these lines were demolished, while roads and paths were no longer passable.

These examples of this form of destruction, which in a greater or less degree accompanies all large earthquakes, suffice to show the danger that is incurred by using these lines in the construction of railroads, the laying of

water-pipes, or sites for construction generally. In some
cases the river banks were made level with the surrounding country, while in others they were raised above it,
having slopes of from three to one to two to one, and
being twenty or thirty feet wide on the top.

Fig. 34.—End of the Nagara River Railway Bridge (Burton)

Raised embankments, like those by which a railway
rises to a bridge, were usually entirely demolished, and
the rails held together by their fishplates were left in mid-
air to hold up the sleepers. The destruction of these
embankments was largely due to the non-coherence of the
sandy material out of which they were built, and which

was shaken down much in the same way that a little pile of sand standing on a plate would subside if the plate were shaken (fig. 34).

The practical lesson to be learned by those who by necessity are compelled to construct in places like those last mentioned is to employ methods and to use materials which give a greater coherence than is demanded in ordinary practice.

Foundations.—In nearly all countries where there is legislation respecting the character of buildings which may be erected in seismic districts, attention is paid to the character of the foundations that are to be employed.

The Ischian regulations provide that buildings must be founded on the most solid ground.

If, however, the ground is soft, a platform of masonry or cement should be formed, which for a one-storey building must be ·70 metre thick, and for a two-storey building 1·20 metres thick. This platform must extend from 1 to 1·50 metres beyond the base of the building. In Manila it is stipulated that the foundations must be able to bear at least twice the weight that is to be placed upon them. When the soil is bad it must be piled or consolidated by a bed of hydraulic concrete, and the foundation of a building must as far as possible be made continuous. These rules, which are based upon the result of experience and experiment, indicate that a building which stands upon a continuous foundation sufficiently well bound together to move as a whole suffers less racking than if it rose from a base the different parts of which might be simultaneously moved in several directions. Several interesting results, indirectly due to non-continuity in foundations, were seen in certain railway bridges after the great Japanese earthquake of 1891.

At the railway bridge over the River Kiso, in central Japan, which is 1,800 feet in length, the piers, which carried 200 feet spans of iron girders, rose from two circular drum curbs, each twelve feet in diameter. The mean height of these piers from the river bed was thirty-

five feet, and the cross-section above the drum curbs was
twenty-six feet by ten feet. The foundations rising from
these curbs were connected by one arch, and it was
through this arch that fracture and lateral displacement
took place (fig. 35).

FIG. 35.—KISO SAWA RAILWAY BRIDGE (BURTON)

Although it seems possible that a differential motion
between the two supporting curbs may have played a part
in causing the destruction, it is more likely that the
principal reason was the fact that the arch created a line
of weakness, and it is along this very line, as will

be shown later, that the greatest strength is required. Foundations of this nature are now no longer used in Japan, and at the Kiso (and other rivers) they have been replaced by elliptical curbs of thirty feet by twelve feet. Since in pits varying between ten and twenty feet in depth the movement recorded is somewhat less than that recorded upon the neighbouring surface, it may be concluded that a building rising freely from a deep foundation, as in the case of a house with an open area and a basement, will be subjected to less movement than a building rising directly from the surface. The practical realisation of this supposition, which was only arrived at after a series of experiments lasting several years, may be seen in several large buildings in Tokyo, forming part of the Imperial University. These have successfully resisted the effects of several very heavy shakings, while neighbouring buildings equally strong, so far as masonry is concerned, which rose from the surface of the ground have been cracked or so far shattered that in part they have required rebuilding.

The most remarkable illustration of this description occurred in 1891, when Tokyo was rocked to and fro by a series of large undulations which originated at a point about 200 miles distant. On this occasion the Engineering College at the University received not a single crack, while its workshop, some twenty yards distant, was so far damaged that it had to be re-roofed and its height reduced.

For local disturbances accompanied by much vertical motion it is not likely that this system of construction would be of much avail, all that is gained by it being to minimise the quantity of motion received from more or less undulatory motion, which chiefly disturbs the surface of the ground. That at least there is no harm in such a method is attested by the fact that in earthquake countries where there is legislation respecting building, cellars or basements are recognised as admissible, while in these cellars vaulting is allowed. For storeys above the ground, arched construction has invariably been suppressed.

Another method of minimising the quantity of motion received by a building is to give it free foundations.

As an example of this I may mention a bedroom attached to my house, which rested at each of its pillar-like foundations upon a layer of $\frac{1}{4}$-inch cast-iron shot placed between two flat iron plates. When first put up, now many years ago, it rested on large iron shells carried in dish-like castings. But it was so unstable at the time of high winds, rolling in an unpleasant manner from side to side, that the shells were replaced by iron balls, about one inch in diameter. Even with these at the time of typhoons the movements were excessive, and the balls were replaced by the shot. The result of this was to introduce sufficient friction to resist the effects of winds, while it was insufficient to overcome the inertia of the building, which therefore tended to remain at rest while the ground beneath rapidly moved to and fro. It is not unlikely that short rollers placed at right angles, or even a layer of rounded pebbles, might be equally effective. This illustration is not brought forward as an example to be followed in practice, but only as an illustration of a principle that has several applications.

On loose or soft ground it might possibly be used with advantage for small light buildings, but there would be difficulty in its adoption for all heavy structures. Also it is ineffective in resisting vertical displacements, or the effects of rolling which accompanies strong undulatory motion.

The first to propose free foundations of this description was Mr. David Stevenson ('Trans. Soc. of Arts, Scot.,' 1868, vol. vii., page 557), who arranged ball joints for two iron lighthouses to be erected in Japan. On their way out from England these were unfortunately lost at sea. A seismic table to carry the lamps at lighthouses was, however, installed at several stations, the behaviour of which we have had many opportunities of studying.

Mr. R. Henry Brunton, who was entrusted with the erection of some of the Japanese lighthouses, gives an example where the chimneys of lamps on one of these aseismic tables were pitched off by an earthquake. In Mr. Brunton's paper on ' The Japan Lights,' [1] it is stated that, after erection, the free motion of the tables occasioned so much inconvenience that the European engineers then in the Japanese service had them clamped, and the arrangement was not adopted in lighthouses subsequently erected. The author, in 1889, learned from the officials in the lighthouse department in Japan that ' in 1882, wishing to give Mr. Stevenson's tables another trial, several of them were put in working order. The result was that on March 11, 1882, at Tsurugasaki, a number of the lamp-glasses on the burners were overthrown. Some time afterwards a second shock produced a similar effect. At neighbouring lighthouses, two of which are within eight miles, and not provided with aseismic tables, no damage was sustained. The shock of March 11 was felt for at least 300 miles along the coast, and its effects at Yokohama and Tokyo, which are at no great distance from Tsuru-gasaki, were carefully recorded. I am not aware that any small articles like lampglasses, bottles, vases, &c., in ordinary houses were overthrown. The fact that no ill effects occurred at other lighthouses provided with Messrs. Stevenson's tables, like those in the Inland Sea and near Kiushu, must not be regarded as an argument favourable to the tables, inasmuch as the earthquake referred to was not felt in those districts. It may here be remarked that one result of the general seismic survey of Japan shows that aseismic tables are no more required in certain portions of the empire than they are required in England.'

As a further illustration of the manner in which aseismic tables have behaved, the author quotes the following translation of a report from the Chief Lightkeeper at Tsurugasaki :

' Sir,—On October 15, 1884, at 4.16 A.M., very severe shocks of earthquake were felt. The aseismic table was in working order, but the shocks were so violent that

[1] *Minutes of Proceedings Inst. C.E.,* vol. xlvii. p. 1.

fifteen lampglasses out of the twenty-one in use were upset and broken. The lamps thus stripped of glasses began to smoke. The milled heads of the wick-holders being shaken off, and besides the revolving machine being in motion, we had some difficulty in replacing the glasses promptly; however, we managed to put them all in proper order again by 4.21 A.M.—I am, Sir, your obedient servant, &c. &c.'

Although these illustrations apparently condemn the seismic joints as used by Mr. Stevenson, it must be remembered, first, that their failure was in certain instances due to having been subjected to strong vertical motion, the effects of which no joint of this description is able to mitigate, and secondly, that, as in the author's first experiments, they allowed of too much freedom.

The only instance in which a house resting on balls several inches in diameter has behaved like a building resting on ordinary foundations is one that was erected several years ago by Dr. W. Van der Heyden in the grounds of the General Hospital at Yokohama.

The reasons that this structure [1] has proved to be satisfactory are twofold. First, the balls rest in deep cup-shaped depressions; and, secondly, the building which rests upon them is, for its size, exceedingly heavy. The combination of these conditions results in considerable resistance to lateral displacement.

An unintentional form of aseismic arrangement is

[1] The walls of this building are made of a series of iron frames, each of which carries two layers of thick glass about six inches apart, the space between them being filled with a saturated solution of a soluble salt. When the sun shines upon the face of these walls, so much of its heat is absorbed in rendering more of the salt soluble that the temperature in the interior of the house is but slightly changed. During the night, by partial crystallisation of the solution, heat is given out, and the rate at which the temperature of the interior falls is greatly reduced. The general result is that, without the aid of artificial heat, a fairly equable temperature is obtained. A second feature of the building is that all air with which it is supplied is cooled and filtered from dust and germs. These attempts to mitigate the effects of earthquakes, avoid changes in temperature, and to render putrefaction less rapid, give to Dr. Van der Heyden's experiment an unusual interest.

found in ordinary Japanese frame buildings, the sills of
which rest loosely upon the upper surface of stones or
boulders planted in the soil.

From experience we know that houses of this description suffer less destruction than common masonry

Fig. 36.—Shonai, N. Japan, 1893 (Omori)

structures, but to what extent this is due to the free
foundation cutting off the motion imparted by the moving
ground is not known. Their relative immunity from
destruction is also dependent upon other peculiarities in
structure, some of which will be referred to in the next
section.

Roofs.—When a building with a heavy roof is suddenly moved forwards, the roof by its inertia tends to remain at rest (fig. 36). The result of this is that the walls pass beyond the tie beams of the frames and there is a collapse, or else there is a tendency to cause a fracture

FIG. 37.—MANILA, JULY 1880

between the lower parts of the walls, which have moved quickly, and the upper parts, which, being constrained by the superincumbent load, have not sensibly altered their position (fig. 37). The damage resulting from the former of these actions may be minimised by allowing tie beams, resting upon wall-plates, to extend across the whole width of the walls, or at least to reach two-thirds across their thickness. The second form of disaster may be mitigated by using light roofs, and by giving them a certain freedom. Certain roofs which are of considerable span in the old Engineering School (Kobudaigako), Tokyo, were built so that they rested freely upon the supporting walls, and were not carried with them in horizontal displacements. Although during the last twenty years they have experienced many severe shakings, hitherto they have remained uninjured.

In Japan we find that temples and other large buildings with heavy roofs have beneath the supporting timbers and the superstructure a multiplicity of timber joints forming corbel-work, which at the time of an earthquake yields, and therefore does not communicate the whole of the motion from the parts below to those above.

In the great earthquake of Ansei, 1855, so far as I am aware, the whole of these buildings remained intact.

In 1891, however, although it was seen that temples stood better than other buildings, still in the epifocal region many of them fell.

Roof trusses should be light and rigid, and for spans exceeding twenty feet the use of iron has been recommended. Most certainly they should not rest immediately above points of weakness such as may be formed by openings in the supporting walls, but should be carried on wall-plates. The roof itself should not be too steeply pitched, it being a common experience that such roofs may lose their covering of tiles or slates, while the coverings of neighbouring buildings with flatter roofs have not been disturbed.

If tiles are used on a roof because they are heavy it is necessary that they should be properly secured, especially near the eaves and along the ridges. In some regulations where tiles have been admitted it is specified that in such cases there shall be above the ceiling a floor of planks, but even then tiles have been allowed only for buildings not more than one storey in height which are not habitations.

Iron, zinc, and felt have been recommended as covering materials. As has been well illustrated in Manila, the difficulties with roofs made of sheet metal are to secure them so that they shall not be disturbed during severe gales, and to protect the interior of the house against heat. In Manila, where typhoons are frequent and the heat is great, the first end is attained by a special system of bolting, while the latter is attained by two or even three false ceilings.

The Ischian law does not prohibit the use of *terrazzo* or flat roofs, but it provides that the framing of the same shall be strong and covered with materials that are fairly light. The Commission on whose reports these regulations were founded condemned such roofs. (For building regulations in earthquake countries, see 'Trans. Seis. Soc.' vol. xiv.)

Walls and Columnar Structures like Chimneys and Piers and for Bridges.—Walls, like chimneys, should be light and strong. If heavy, and especially if loaded in their upper parts by copings and balustrades, they may be fractured and shattered by their own inertia. The height to which walls may be taken with safety depends upon the material of which they are constructed, the nature of the roof, &c. In Ischia it was suggested to limit buildings to two storeys, or a height of 7·5 metres (24·6 feet). The regulations, however, give 10 metres (32·8 feet) as a limiting height, and they must be of simple masonry or tuff to a height of 4 metres (13·12 feet), with a thickness of 0·70 metre (2·30 feet). The committee suggested that external walls should be at least 0·30 metre (0·98 foot) in thickness, and that their uniformity should in no way be broken by openings for chimneys, pipes, &c.

The Ligurian regulations allow three storeys above the cellar, and a height of 15 metres (49·2 feet). The walls, if not built on the barrack system, should be at least 60 cm. (23·6 inches) thick, and have a batter one-twentieth of their height. The Norcian regulations allow two storeys above the cellar, and a height of 8·5

metres (27·88 feet). If a third storey existed it was to
be destroyed. The walls were to be thicker than ordinary,
and their thickness was to vary with the material employed and the height of the structure.

In Manila masonry walls of ordinary dwellings only
reach the first storey, the upper storey being of timber. The
walls for public buildings, however, may be higher. The
regulations specify that the upper walls must not rest on
a floor.

The length of a wall should not exceed twice its height
unless supported by a buttress. Such buttresses might be
placed at intervals not greater than twice the height of a
wall. Its thickness must be one-fifth of its height. Outside
walls, transverse walls, and buttresses must be well united,
while the corners of buildings should be supported by
buttresses.

It would appear that the system of building with an
upper storey of wood resting on, and not built into, the
supporting wall, and a light roof, ought to do much
towards insuring the stability of a building. The weight
of ordinary masonry may be reduced by the adoption of
hollow bricks.

A very important point not to be overlooked is to give
the walls forming the ends of a long line of buildings
additional strength, it being not an uncommon experience
to find that the end wall of the last house in a row has
been shot forwards like the last truck on an uncoupled
train which has been bumped at its other end by a
locomotive. Walls made of stone or brick as a facing to
a heavy internal wooden framing often seem to suffer
badly, the latter by its movement entirely destroying its
outside covering. In a similar manner a covering of tiles
on the face of a wooden building may be shaken down,
and the same remark applies even to a heavy covering of
plaster. If it is necessary to face a frame building with
any materials, unusual security must be given to their
attachment.

Experiment has shown that when a column standing

vertically on a truck to which it is firmly attached is moved quickly back and forth, there is a certain rapidity of motion which causes the column to be fractured at or near its base. A still quicker motion is necessary to cause the broken column to fall. From a diagram of the motion of the truck the acceleration experienced at the time of fracture has been calculated, and the following formulæ obtained.

Calling this acceleration a, then the relationship of this to the dimensions and the strength of the column may be expressed as follows :

$$a = \frac{1}{6} \frac{gFAB}{fW}$$

where

    $F =$ the force of cohesion, or force per unit surface, which, when gradually applied, is sufficient to procure fracture.

    $A =$ area of base fractured.

    $B =$ the thickness of the column.

    $f =$ height of centre of gravity of the column above the fractured base.

$W =$ the weight of the portion broken off.

The quantity F, which was determined in a testing machine varied in the columns which were broken between 4 lb. and 14·8 lb. per square inch, corresponding to which different values for a were obtained. Out of fourteen experiments, the values obtained for a in twelve cases were fairly comparable with the values for the maximum accelerations calculated from the diagrams of motion. From this formula a second formula, showing the height to which a wall may be built capable of resisting an assumed acceleration, was obtained.

It may be written $x = \sqrt{\dfrac{FBg}{3aw}}$, in which $x =$ the height of the wall and w the weight of a cubic inch of brickwork, or ·0608 lb.

M

From this last formula, assuming the greatest acceleration we may expect in a district to be, say, 1,000 mm.
per second per second, we can determine the height to
which a column or wall of given section, and made of
materials with a known tensile strength, may be carried
above its foundations, and be just able to withstand the
given motion. For example, a column of brick two feet
square, having a value for F of 15 lb. per square inch,
would just be on the point of fracturing if it was eleven
feet seven inches in height.

In the practical application of this formula it must be
remembered that it is an easy matter to obtain a very
much larger value for the strength, and a much less value
for the weight, of the materials employed than those
taken in the illustration.

Further than this, it is not necessary that the wall
or column should be of uniform section from base to
summit. After an actual earthquake the fracturing at the
base of columnar structures is often very pronounced, being
well illustrated in the Japan earthquake of 1891 by a series
of piers carrying a railway bridge across a river.

If the piers are of cast iron, circular in section, and
filled with concrete, a few of the shorter of those near the
river bank may stand intact; a short distance out, where
they are longer, all will be fractured near the base ; while
near the centre of the river, where the columns are taller
but still of the same section, not only will they be fractured
at their base, but they may be overthrown and broken into
fragments by their fall. It may here be remarked that
structures of this description have in Japan been entirely
replaced by masonry. With the masonry, although the
destruction is not so great as it is with cast iron, it nevertheless is similarly distributed ; the short piers near the
banks may stand, while the longer piers in the centre of
the river may be fractured at their junction with their
foundations.

Inasmuch as the short piers which have withstood a
shaking have approximately the same general form and

cross-section as the taller piers which have failed, the idea
suggested itself that if a portion of the material used in
the construction of the former had been added to the
latter, all destruction might have been avoided.

A partial illustration of the distribution of destruction
was shown by the varying amount of fracture and displace-

FIG. 38.—RAILWAY BRIDGE OVER THE RIVER NAGARA (BURTON)

ment shown in the piers of the bridge crossing the River
Kiso, to which reference has been made.

A similar illustration was to be seen in the piers of the
railway bridge crossing the River Nagara (fig. 38). These
piers were composed of groups of hollow cast-iron columns,

each two feet six inches in diameter and one inch thick, and filled with concrete (fig. 39). Five of these columns, strongly braced together, formed a pier, which carried about 185

Fig. 39.—River Nagara Railway Bridge (Burton)

tons. Although the average height of these piers above the river bed was twenty feet, those near or on the shore of the river were very much shorter than those in the middle of the river. After the earthquake of 1891 the central part of the bridge totally collapsed, whilst to the right and left of the ruin, although the cast-iron columns were fractured, the amount of destruction was less, and on the banks the piers remained upright (figs. 38 and 39).

It may be mentioned that these bridges had repeatedly withstood the effects of dangerous floods, and the effects of typhoons which had overturned locomotives. To break up the piers of Kiso, preparatory to rebuilding, it was necessary to employ dynamite. Since yielding first shows itself at the base of a column, it is evidently necessary that its lower section should be of greater dimensions, or be built of stronger materials, than that which comes above it, or, to extend our idea, every horizontal section of the structure should be sufficiently strong to resist the effects of the inertia of its superstructure.

In a square column, for example, where $x =$ half the dimensions of any given section whose distance from the top is y, then for a of our original equation to remain constant,

$$y^2 = 10 \frac{gF}{aw} x.$$

If the column has a circular section,

$$y^2 = 7\tfrac{1}{2} \frac{gF}{aw} x.$$

If it is rectangular,

$$y^2 = 4 \frac{gF}{aw} x.$$

From this it is seen that, with the same dimensions of base and height, the strongest column is the one with a square section.

As an illustration, for a column with a square section suppose

$a = 1,000$ mm. per second per second,
$F = 5$ lb. per square inch,
$w = 0.0608$ lb. per cubic inch.

Then with x and y expressed in inches,

$$y^2 = 8,100x.$$

The outline of this column, which is parabolic in form, is shown in fig. 40.

In practice its upper portion would naturally be cut off, while its sides would be stepped or run up in straight lines. Very many piers following these laws have been built on the Usui Railway in Japan, by Mr. C. A. W. Pownall, M.Inst.C.E., who, in addition to approximating to the required form, also obtained greater security by using a stronger cement in the lower portions of the piers than in the higher (fig. 41). A building with walls of a form approximating to the formulæ has been erected by Professor Tatsuno, and is used as a small observatory in the University grounds. Although this building forms an excellent object-lesson, on account of the excessive use of materials it requires, it is not to be recommended as a type of structure for an ordinary dwelling.

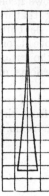

Fig. 40

The next class of structures to be considered are tall chimneys, which in certain towns in Japan, like Osaka, are already covering the temples with a canopy of smoke and giving to the city a Sheffield-like appearance. These are circular or rectangular in section, slightly tapering, and occasionally buttressed on the outside. Whatever their form may be, when well shaken, after waving back and forth through a distance of several feet, they nearly all fracture to a great extent vertically, and then collapse at about two-thirds their height.

To be on an eminence at the time of a strong earthquake, as many of my friends have been, and see tall chimneys swinging until they fall to join the ruins of the smaller chimneys from the dwelling-houses, is a sight

which I never witnessed. All that I have seen are the
results, and the similarity of these is very striking.

FIG. 41.—BRICK PIERS IN THE USUI RAILWAY (KIK KAWA)

Some builders have strengthened the weak section of
their chimneys with hoops of iron, but it is not likely that
such support will materially mitigate disasters. After the
earthquake of 1890 three chimneys which stood amidst

their ruined neighbours, and are therefore deserving of notice, were constructed by Mr. J. Diack, of Yokohama, who, rather than looking to obtain rigidity by means of hoops of iron, obtained a transverse elasticity by an ingenious system of longitudinal iron bonding. In many instances masonry has been discarded in favour of iron, and where this material has not been entirely employed, it has been used for the upper third or half of a chimney.

At Kanegafuchi cotton mill, in Tokyo, there is an iron chimney of about 150 feet in height, with a diameter at the base of about twelve or fifteen feet. It is lined with bricks, and receives additional support from radiating ties of iron rods. On June 20, 1893, when most of the tall chimneys in Tokyo were fractured, all that happened at Kanegafuchi was the snapping of one of these ties.

If masonry has to be employed, the only escape from disaster, over and above the system of construction introduced by Mr. Diack, appears to be the application of the principles formulated for the piers of bridges, namely, the reduction of top weight, the use of hollow bricks, and giving to the weak section a strength greater than that which has hitherto been considered necessary.

An engineering acquaintance of mine who was censured for the badness of the mortar he had employed for a tall chimney that had totally collapsed, defended himself by the remark that had he used a better quality of cement, the chimney, instead of falling as a heap round its foundation, might have toppled over and destroyed the neighbouring houses.

The defence, although characterised by a naiveness almost amusing, suggests the idea that, whenever possible, tall chimneys should be located so that if they must fall they shall create the least amount of ruin. With householders, after an earthquake, it is not the loss of their chimneys which they regret so much as the loss the chimney has caused by crashing through a roof and several floors.

That a tall chimney may stand better by itself than

when tied in anyway to a neighbouring building was well
illustrated at the Yokohama Iron Works in 1880, when
a chimney was cleanly cut in two by an iron band which
had been placed round it and tied to a neighbouring
building. By itself the chimney might have stood, but in
consequence of its movements not synchronising with
those of the building to which it was fastened it was
destroyed. This last illustration, showing how destruction
may occur in consequence of a nonsynchronism in vibra-
tional motion of two structures which are connected to
each other, leads to a consideration of the manner in
which the chimneys of ordinary dwellings are destroyed.

It often happens that after an earthquake of moderate
severity almost every house in a town, although it has
not suffered any other appreciable damage, has had its
chimneys shattered, rotated, or shaken down, and the
point at which they yield is almost invariably at their
junction with the roof.

The first writer in Europe who recognised that one
portion of a building might destroy another, in consequence
of a want of synchronism in their movements, was Bertilli,
who referred to the matter in 1887.

In Japan the same subject had been written about, and
experimented upon, and rules had been adopted to obviate
the cause of destruction as early as 1880. In that year
Yokohama lost many of its chimneys in consequence of the
wooden framing of the houses swinging against them.
Crooked as the chimney of a dwelling house usually is,
it will, if freed from the surrounding building, withstand
a shaking of considerable violence. That this was the case
was seen in a series of exceedingly unstable-looking
chimneys, which in consequence of a fire prior to the
earthquake of 1880 had been left standing. To the
surprise of all who saw them, they remained standing
after the earthquake, which destroyed nearly all the
other chimneys in their vicinity.

The rules regulating the construction of chimneys are
but few. The Ischian law states that they should be

isolated from the walls ; that of Liguria that they should not be in the walls, nor connected with the building, and should be low. Chimneys not being much required in Manila, nothing is said about them. Experience in Japan has taught householders to build their chimneys as short and thick as possible, to allow them to pass freely through the roof, and not to load them with heavy coping stones. After the experiences of 1879 and 1880, many of the residents in Yokohama materially altered the form of their chimneys. In 1887 these buildings did not suffer, the buildings which did suffer being chiefly those put up subsequently to 1880 and without any regard to the experience of previous years.

From this time up to 1894, although occasionally a solitary chimney was shattered, the experiences of 1880 were forgotten, and for the sake of architectural appearances a new crop of chimneys with heavy copings and ornamental tops grew up. On June 20, 1894, these all came down.

Brick shafts are now terminated at the roof, from which they are continued upwards, with an iron super-structure, some of which in form and colour are barely distinguishable from their dangerous brick predecessors. Although it might be argued that a building really supports a chimney up to its weakest point, and that therefore it gives way at its junction with the roof in consequence of its own inertia, which in part may possibly be true, nevertheless the balance of evidence seems to indicate that a chimney and a house may be mutually destructive.

ROTATION OF COLUMNAR STRUCTURES

A form of destruction particularly noticeable in cemeteries is due to the overturning or rotation of tomb-stones and monuments, to obviate which it is necessary to allow each stone column to rise from a socket cut in its pedestal. If there are series of stones one above the other, these are connected by bonds and dowels. That the rotation of stone columns or a fractured chimney top does not necessarily imply any rotational motion of the earth, but may be due to a rectilinear movement, was first demonstrated by Professor T. Gray. If a light but tall rectangular box is placed on a table to which a rectilinear vibratory motion can be given, it will be observed that there will be no rotation if one face of the box is at right angles to the direction of the motion. If this face be placed obliquely to the direction of motion, then there will be a rotation, and the direction of rotation will vary with the degree of obliquity. Thus, in the figure representing the plan of a column, a back and forth motion in the direction $a\,a$ or $b\,b$ will only cause the box to rock on its edges. A similar motion in the direction $c\,c$ will also cause the box to rock on its corners, and theoretically there should be no rotation.

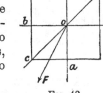

Fig. 42

A continued movement in any direction between $a\,a$ and $c\,c$ will cause right-handed rotation, while if it be between $b\,b$ and $c\,c$, the rotation will be left-handed (fig. 42).

The explanation lies in the fact that the force due to the inertia of the column, acting in any direction o F opposite to that of the direction of shock, may be resolved into two forces at right angles to each other, one along $o\,c$ tending to tilt the column on its corner, and the other to turn it round. After the earthquake of 1880 the writer observed that parallel rows of similar columns in the Yokohama cemetery rotated in the same direction, and that direction of rotation therefore indicates possible direction of movement.

CHAPTER IX

EARTHQUAKES AND CONSTRUCTION (*continued*)

Connection of different parts of a building—Buildings in San Francisco
—Lescasse system—Temple roofs—Floors—Archwork and wing
walls—Doors and windows—Lines of weakness—Balconies, cornices,
gables, ceilings, and staircases—Materials—Form of Bricks—Types
of buildings—Earthquake lamps—The barrack' system—Systems
of building in various countries—Construction underground—
Reservoirs—Water towers—Conclusions relating to building—Sea
waves.

CONNECTIONS BETWEEN DIFFERENT PORTIONS
OF A BUILDING

SINCE time immemorial, buildings have been tied together
with iron or with wooden rods; but some time previous
to 1868, when San Francisco was shaken, a patent known
as the Foye patent was taken out to improve the construction of sea walls. This was made to apply to land
structures. The City Hall and other buildings in San
Francisco are built upon this plan, which consists in tying
together the walls at each floor by transverse and fore-
and-aft rods of steel or iron. A plan similar to this is
that of Mr. J. Lescasse.[1] It has been applied to several
buildings in Tokyo and Yokohama.

For such earthquakes as these buildings have experienced—excepting on one occasion, when the chimneys of
the German Hospital in Yokohama were more or less
injured—they have stood well. This system, however,
requires to be thoroughly executed; for if the rods be

[1] *Mémoirs de la Société des Ingénieurs Civils*, 1887, p. 212.

too few, or if the bearing surfaces be too small, rather than support a building they accelerate its destruction, especially at the points of contact. Such buildings, partly for this reason and partly on account of their expense, are not looked upon with favour in Italy. The Ischian law specifies that if iron bands or chains are used they must act upon a large surface.

The Tokyo disturbance of June 20, 1894, which produced disastrous results among very many ordinary European buildings, does not appear to have produced any visible effects upon several buildings in which the Lescasse system had been adopted.

Instead of tying a building together until it may have a rigidity which may be likened to a steel box, the builder may go to the opposite extreme, giving all his connections so much freedom that each part of the structure may be capable of yielding in the same manner as a wicker basket. That temple roofs in Japan probably stand in consequence of the freedom existing between them and the supporting walls has already been pointed out, and there is no doubt that Japanese dwellings owe much of their security to the freedom with which they yield. Their weakness chiefly resides in their heavy roofs, and the reckless manner in which main supports are cut away at joints with other timbers.

After the earthquake of 1893 the visitor to the provinces of Mino and Owara saw illustrations of destruction due to the failure of vertical supports in the apparently triangularly formed thatched roofs of the farmers' houses, which dotted the country in all directions, looking like so many huge saddles. With lighter superstructures and iron straps and sockets to take the place of many mortises and tenons, there is but little doubt but that the destruction of life and property would be mitigated.

An excellent example of a large and handsome building put up on these principles is seen in the Imperial Hotel in Tokyo, which, amongst its other attractions, is advertised as being earthquake-proof. Although it would be exceedingly difficult to demolish such a structure by shaking, its easy yielding results in so much internal disfigurement by the cracks in plaster and the falling of portions of ceilings and the like that its advantages are somewhat marred ; guests may be needlessly alarmed, and the building lose its reputation.

FLOORS

When building to obtain rigidity, much may be done by paying attention to the floors. The beams supporting one floor should be placed at right angles to those on the floors above and below, and all should extend nearly, if not completely, through the supporting wall. Floor joists should similarly be well supported at their extremities, and, if possible, cross each other at right angles, while the flooring should be laid diagonally. Experience has shown that much destruction has been occasioned by the withdrawal of beams and joists from their supports, and both in the Ischian and Nortian edicts relating to these matters the construction of floors receives special attention.

ARCHWORK AND WING WALLS

An ordinary arch is undoubtedly stable for vertically applied forces, but for horizontal stresses it is one of the most unstable structures that could be erected. So often has the arch been the cause of ruin when shaken by an earthquake, that special rules have been drawn up in Italy and Manila respecting such structures. Thus, in Manila intersecting vaults are not allowed, and ordinary vaults are only permissible when strengthened in a particular manner by iron. In Liguria vaults can only be used in cellars, and even there the rise must be at least one-third of the span. The law of Norcia also permits the use of arches in cellars only, and their thickness and the method of construction are defined. In Ischia archwork, with a

rise of one-third of the span and with a thickness of
·26 m. at the crown, may be used, but only in cellars.
Speaking generally, the use of archwork above ground
has been prohibited, and if it has existed after an earthquake, all Governments who have paid attention to

FIG. 43.—RAILWAY VIADUCT, JAPAN, 1891 (BURTON)

building have ordered its removal.   Underground its use
is permitted providing that the arches are not too flat.
This, however, only tells us that the motion beneath the
surface is too small to destroy even a bad form of structure,
and, therefore, such a form of structure, providing it is
underground, is allowable.

In 1890 the brick arches of railway bridges apparently collapsed by the outward movement of the abutments, which in consequence of the arch cracking at its crown and then falling inwards, like a toggle joint, resulted in their being forced still further backwards (see fig. 43).

If it is a necessity for arches to exist, they should not be too flat; they should have a specified thickness, be protected by an iron or wooden beam above, and curve into their abutments. Arches which meet their abutments at an angle often show cracks at their junction, and these may have been formed by very slight shakings. Light arches connecting heavy walls, or arches for porticoes, supported on one side by a building and on the other by a column, often give way. Wing walls, such as support an embankment and form an entrance to a subway beneath a railway, as usually constructed seem to be especially weak.

In 1891, in Mino and Owari, I do not remember having seen a single wing wall which had not separated along a vertical line from the abutments of the bridge.

Openings in Walls. Doors and Windows

In the building regulations for Norcia and Ischia it is stated that openings should be placed vertically above each other. It appears to the writer that if we have a series of openings like doors and windows in a wall placed vertically above each other, it is very much the same as if we had here and there built our wall with the joints of a line of bricks or stone continuously above each other—that is to say, we have destroyed the uniformity of the wall by lines of weakness which will readily give way to horizontally applied stresses.

Although the subject may not be one of great importance for ordinary dwellings, the writer inclines to the opinion that the doors and windows in successive tiers of openings ought *not* to be above each other, but as far as possible arranged so that any line of openings, when regarded vertically, is as much broken as possible.

FIG. 44.—CHURCH AT MANILA, 1880

After an earthquake we often meet with buildings which have been disfigured by lines of cracks running vertically downwards from window to window, these openings having performed a similar function to the perforations in a sheet

N

of postage stamps. This is illustrated in fig. 44, which
shows a vertical line of yielding in a church in Manila
which was partially shattered by the earthquake of
July 1880. To arrange doors and windows so that
they may form ready means of escape is certainly a
matter worthy of attention. An important point mentioned in the Ischian law is the position of doors and
windows relatively to the freely vibrating end of a building, the limiting distance being 1·50 m. Similar regulations exist in Norcia and Liguria. This distance should,
if possible, be made to depend upon the materials of which
a wall may be constructed, its dimensions, and the size
of the openings.

A terrible destruction of life has often happened in
consequence of the openings on one side of a row of
houses, as, for example, in shop fronts facing a street,
having been much greater than those in the opposite side.
This is particularly marked in Japan, where, no matter
from which side the shaking may approach, the tendency
of the buildings is to yield on their weakest side and fall
inwards upon the streets. If these are narrow, the *débris*
from the two sides forms an embankment down the middle,
beneath which the inhabitants seeking refuge from their
houses are entombed.

BALCONIES, CORNICES, GABLES, CEILINGS, AND STAIRCASES

In Ischia it was suggested that balconies should not
project more than ·60 m. beyond the wall, and should
be so constructed as to form a part of the wall.
The regulations provide that cornices should not project
more than ·30 m. beyond a wall. From the Ligurian
regulations we learn that cornices shall not project beyond
the thickness of the wall to which they are attached, while
roofs may not rest upon them. Stone consols must run
through the wall to which they are attached. In Manila
the regulations require that the balconies rest on the

prolongation of timbers of the upper floor. Otherwise a special form of construction is required. From what I saw of the balconies or upper verandahs when in Manila, it appeared that many of them were without support on their outer sides. In such instances they act as loaded cantilevers, which, either for horizontal or vertical motions of the building, must cause considerable stress at their points of junction with the supporting wall. A careful examination of several hundreds of brick houses in Tokyo showed that the walls were usually cracked at the points where they were entered by the beams supporting a balcony, notwithstanding the fact that the same balconies were supported along their outer face by vertical pillars rising from the ground. My own opinion is that balconies in any form are objectionable features in a building constructed to withstand earthquakes.

Walls which are run upwards beyond the height of those carrying the main roof so as to form gables, especially when they support a heavy coping, are not referred to in regulations, but they certainly form dangerous adjuncts to a building, and usually fall outwards, destroying and burying porticoes, or whatever may be beneath them.

Staircases are also overlooked in regulations, but if heavy and supported from the walls, they may have the same destructive cantilever action that is possessed by balconies. Ceilings, we are told, should be constructed in the ordinary manner with laths and plaster. My own observations suggest that this is by no means sufficient. The laths should be well secured, and the plaster be especially adhesive. Heavy ornamentation should be avoided. Although persons are not likely to be killed by the coming down of a few pieces of plaster, the cracking and falling of it cause so much disfigurement and alarm that even a building which has been constructed to be earthquake-proof may receive a bad name, and be shunned by those who have a choice in the matter. In the case of a large hotel the falling of plaster may mean financial ruin.

N 2

MATERIALS

This section, relating to the quality of builing material which ought to be employed in earthquake countries, is one which cannot be too greatly emphasised.

All regulations relating to this matter insist that material of good quality should be used.

The Ischian regulations specify that for the principal framework of buildings chestnut must be used. In all cases squared stones are to be employed. The lime must be good, and be properly slaked with fresh water. Ground hydraulic mortar must be used below, and the sand for the mortar must be clean. These matters are dealt with in all regulations. In the regulations for Manila there are special remarks condemning the use of liquid lime, and recommending that stone walls should be kept wet while the mortar is setting, also that there should be a good bonding, &c.

After the Japanese disaster of 1891 I had occasion to test the tensile strength of very many samples of masonry collected from the ruins throughout the earthquake district. In some instances the brickwork used in certain buildings was held together by a material which, although it looked like mortar, was so non-adhesive that the bricks could be easily separated by hand. The tensile strength of this material was too low to be measurable. In Tokyo a tall chimney and a wall, which were built with what appeared to be a similar material, were overturned by a high wind. The strength of other samples varied between 3 lb. and 15 lb. per square inch. When a brick-work structure is shattered, it will be noticed that either the cementing material has yielded or has separated at its junction with the bricks, or that the bricks themselves have fractured. There has been, in fact, a want in uniformity of strength throughout the building.

To obtain economically the uniformity which is desirable, it would seem that the strength and adhesiveness of a cementing material should be approximately equal to

the strength of the material it is intended to hold together,
or, especially for the higher parts of the building, it should
at least have a resistance equal to the effects of the tensile
strain it may have to resist.   To economise the use of
cement in Japan, bricks which lock into each other have
been made (see fig. 45), while hollow bricks have been
employed to avoid effects due to inertia.   In the preceding sections reference has been made to the use of
adhesive materials in embankments, the employment of
iron sockets and straps to avoid excessive cutting in
timber joints, the untrustworthiness of cast iron in

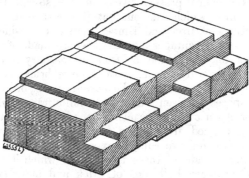

FIG. 45.

columnar structures, and the effects of heavy materials
for roofs.   Materials to resist earthquake effects should
be chosen so as to give a maximum strength with a
minimum weight, and this especially in superstructures.
For example, it would seem that a wall made of pumiceous
scoriæ and cement would be lighter than one where the
cemented basis was ordinary gravel.

Types of Buildings

The type of building most suitable for earthquake
countries was discussed at considerable length by the
Commission summoned after the disaster in Ischia.

The objections to iron buildings chiefly rested in their cost, the difficulty of keeping them cool, and the fact that as they were a novelty it might be difficult to get them generally accepted. The Commission, however, considered them durable and secure, and recommended that experimental buildings should be erected.

Timber buildings, although sufficiently strong and elastic to resist earthquake motion and at the same time fairly impervious to heat, have the disadvantage that they are not durable and are subject to fire. These objections may to some extent be overcome by the proper application of paints, chemical preservatives, and the so-called earthquake lamps, which are put out if overturned. Mixed constructions of iron and timber were not considered to present great advantages over those wholly made of timber. Buildings may be made of iron or masonry either by covering an iron framework with stone or brick, by building an iron framework inside the masonry walls, or by filling up the spaces between a double metallic framework with hollow bricks or other materials. Such buildings, although exceedingly good from many points of view, have the drawback of being exceedingly expensive.

Having considered these types, from which it will be observed ordinary buildings of brick and masonry have been excluded, the committee describe a ' barrack system ' of building, which they particularly recommend for Ischia. Briefly, such a building consists of a timber framework well braced together, the spaces between the timbers being filled up with hollow bricks or some light material like scoriæ. The timbering is hidden by rough-cast. After the disaster of 1755 such a system was made compulsory in Portugal. A building of this type, which may be made ornamental with an outside covering of tiles—which the author, however, does not think is to be recommended—is cheap, impervious to heat, and safe against earthquakes and fires. This suggestion respecting the system of construction was adopted in the regulations issued by the Italian Government.

In the building regulations for Norcia the barrack system is the one to which preference is given. In the Manila regulations considerable latitude is allowed as to the system of construction. Stone walls are considered best, but concrete or brick are also considered good. Although timber offers great resistance to earthquakes, its destructibility by fire, white ants, ordinary rot, and its inability to exclude heat prevent its recommendation. An iron framework filled in with concrete is spoken of with favour. In the recommendations of a committee appointed to consider building in Manila we find that stone is recommended for the basement and for the walls of the ground floor. This, with an upper storey of timber, is the type of building common in Manila (fig. 37).

The military committee which was summoned in connection with the destruction in Manila in 1863 pointed out that destruction had occurred in all classes of buildings, but that buildings with masonry supports had suffered more than others. This led them to suggest that only one kind of material should be used in constructions, and masonry supports should be avoided. Private buildings should be of wood. In all cases the limiting spans of roofs were specified, and the roofs must be light. Lieut.-Colonel Cortés, who wrote at some length on structures in earthquake countries, shows that buildings must be light as well as strong, and this may be obtained by building their parts together much in the same manner that the timbers of a ship are bound together. Foundations and walls should be continuous. Timberwork and masonry should not come in contact, otherwise they may be mutually destructive.

After criticising the system of building in Manila, and showing how it may be improved, especially with regard to balconies and roofs, Colonel Cortés proposes, as a foundation, a timber platform almost on the surface of the ground, from which rises a building with iron or timber framing footed on a plinth of masonry, and surmounted by a light roof. The wall framing may be filled with brick or plaster. Colonel Cortés's descriptions are accompanied by an elaborate series of illustrations.

The Californian system of construction—for which a patent has been granted, as we have said—appears to be very similar to that proposed by Mr. Lescasse, the essential feature being to tie a masonry construction together at each storey by a set of iron or steel rods, which run from end to end, and from back to front in the interior of the walls of a building. There are also rods running vertically.

From South America but little information has been obtained. In Columbia the smaller houses have been built of thick adobe bricks, while the Spanish have used stone.

In Equador (Quito) a special earthquake-proof room is occasionally built, the walls of which are a wooden framework filled in with adobe. Many houses which have adobe walls three feet thick have only one storey, and there are few houses with more than one upper storey.

In Venezuela, also, the houses are low. In Mexico and Bolivia the houses are solidly built; while in Lima certain buildings are constructed lightly, so that they may yield.

From Guatemala (San Salvador) I received from Messrs. Clark & Co., contractors, the drawing of a house supposed to be earthquake proof. It is of timber well framed together, and very similar to the bungalows in Japan. These descriptions from South America are particularly meagre. For a full description of the system to which they refer, see vol. xiv. of 'Trans. Seis. Soc. of Japan.'

CONSTRUCTION UNDERGROUND

It has already been mentioned that at a short distance beneath the surface earthquake movements are somewhat reduced. Nevertheless in certain localities—as, for example, along lines parallel and near to free surfaces and above water-bearing strata—danger may be anticipated.

The effect of violent compression of watery beds is to
cause the ooze, or material of which they are formed, to
force a passage to the surface. From the vent or fissure
thus made, repeated compressions eject sand and other
materials, which accumulate to form cones and ridges. At
the time of the Shonai shock, which occurred in North
Japan in 1893, not only were well tubings shot vertically
upwards, but the wells themselves were filled with sand.

It is obviously important in the laying of pipes to
carry water or sewage, that those lines should be selected
which are the least likely to be interfered with by
actions of this nature. On January 18, 1894, Yokohama
suffered inconvenience by the fractures of the water pipes.

RESERVOIRS

At the time of the Japanese earthquake of 1891, but
distant 200 miles from its origin, I observed waves in a
water tank built in the ground, eighty-two feet in length,
twenty-five feet wide, and twenty feet deep. This tank,
the sides of which are practically vertical, contained about
seventeen feet of water, and the waves ran backwards and
forwards across its breadth, rising first on one side and
then on the other to a height of two feet, splashing to a
height of four feet. Had the walls of the tank risen somewhat above the surface of the ground, and had the motion
been somewhat greater, which it might well have been,
not only would the walls have been ' topped,' but great
pressures would have been applied, which, unless especially provided against, might have resulted in destruction.

Dams impounding waters in valleys are liable to suffer
from similar actions, and therefore require greater heights
and strength than those which have simply to withstand the
steady pressure of quiescent water. All that happened to
the tank mentioned above was a slight separation between
the end and side walls, followed by leakages, but probably
due more to the movement of the ground than to that
of the water. With a curved junction and more batter,

it is likely that such fracturing would not have taken place.

On the occasion of the same earthquake, a partition wall in the service reservoir of the Yokohama Water Works was completely overturned by the back and forth lashing of the water.

Professor W. K. Burton has suggested that the evil effects of waves in ponds and reservoirs might possibly be obviated by a wall or screen rising above the upper level of the dam. As such a screen would have to resist the effects of a moving body of water, it would seem desirable to give its inside face a curvature, so that the impulse of the momentum would be gradually applied.

WATER TOWERS

In an earthquake country the placing of water tanks or the erection of water towers in high buildings is a practice that should be inadmissible. Even the small tanks used for the supply of water to locomotives at railway stations require to be built with unusual care. If simply supported on cast-iron columns, as in 1891, their destruction is certain. A better form of structure would be one following the rules laid down for tall columns, or, in place of that, a framework of timber. Although the fallen gravestones shown on p. 41 illustrate effects which have been referred to in preceding sections of this article, they also show the nature of the destruction which in 1891 overtook so many water tanks. To the constructor the figure illustrates the necessity of extremely free or else extremely rigid attachment to the moving earth, which, in this particular instance, would be best secured by deeper sockets and the use of dowels. Also the centres of inertia should be low.

CONCLUSIONS

If we wish to mitigate the effects of earthquakes, one general conclusion that may be drawn from the present

discussion is to select a site where we know from experience or from experiment that the ground suffers
comparatively slight motion. This will generally be the
hard ground, which is usually the high ground. Soft
ground, slopes, and scarps should be avoided. Having
obtained our site, we can follow one of two general
systems of construction—either to give so much rigidity
to a structure that it may be likened to a steel box, or to
erect a building which is light, but which has so much
flexibility that it may be compared to a wicker basket. In
either of these structures we ought to have lightness,
especially in their upper parts.

Amongst the former class of buildings—which, from the
materials of which they are constructed, are unquestionably
heavy—we have ordinary structures of stone or brick (by
preference we might use hollow bricks). These should
rise from a deep foundation, have a free basement, walls of
unusual thickness, and be well bonded and tied together.
The roofs should be light, and the precautions respecting
the position and form of openings, the arrangement of
floors, roof trusses, and top weight carefully attended to.
In this case we have a building where its strength more
than outweighs the ill effects due to its weight. Such
buildings are durable and relatively safe against fire ;
they are suitable for all climates, but they are relatively
very expensive. The expense limits this type of structure
to buildings of importance.

Light buildings which have sufficient strength and
flexibility to overcome the disadvantages of their own
inertia when shaken by an earthquake are nearly all well
constructed structures of wood or iron. Wooden buildings, however, are neither durable, nor safe against fire,
nor impervious to heat and cold. These objections may,
however, practically be overcome, and their cheapness is
an advantage. Iron buildings are relatively expensive,
and without special arrangements they are too hot in
summer and too cold in winter.

A type of building which is comparable to a brick or

stone structure as regards immunity from fire and behaviour under changes in temperature, but is very much cheaper and at the same time safe against all ordinary earthquakes, is the building constructed on the barrack system, so strongly recommended in Italy. The framing may be of wood or iron, while the filling-in material which forms the walls, which ought to be as light as possible, may consist of hollow bricks or a concrete of light material. For this latter purpose experiments might be made in Japan with a concrete made of the pumiceous light scoriæ, of which there is such an abundance in that country. Ordinary structures in bricks or stone are usually bad, while timber structures with a masonry front are worse. To resist earthquake motion we require lightness, strength, and, if possible, a certain elasticity. Weight, unless it is accompanied by great strength, should be avoided.

For ordinary buildings, unless the barrack system be adopted, I would suggest that for a country like Japan ordinary frame buildings continue to be used. To improve them they require more diagonal bracing, less cutting at the joints, lighter roofs, and some protecting covering against fire.

A type of structure, several examples of which are springing up in Tokyo and which embody several principles promising to render them less liable to destruction than ordinary dwellings, has been recently designed and patented by Mr. Inouyé, a private architect. The main feature is that the principal rafters of the roof are carried downwards to a soleplate on the ground. Instead of mortises and other joints, which have a weakening effect, cast-iron sockets and a variety of iron straps are employed. A verandah gives the appearance of walls, and the roof, which is of wood, is covered with paint and sand. When completed these buildings have a villa-like and pleasing appearance, and in the writer's opinion will stand shocks that will destroy ordinary buildings (fig. 46).

To the engineer and builder who carries on his

occupation in a country like England the direct application of what has been noted in these articles is but small,
yet it is, perhaps, possible that here and there a hint may

Fig. 46.—A New Type of Japanese Dwelling

be derived when designing a machine subject to vibratory
motion, or carrying out a construction intended to resist
the effects of winds and waves and other horizontally applied

stresses. Directly, however, we turn our eyes away from our own islands to our earthquake-shaken colonies, and other countries in which our capital is invested, the question discussed in these pages becomes important, and this imperfect compilation of facts and suggestions derived from experience and experiment may prove of some service. A remarkable illustration of this may be derived from our legation and consular buildings in Tokyo. At first they were ordinary buildings with tile or slate roofs, ornamental copings, tall chimneys with heavy tops, such as we should find in houses with some pretensions in provincial or rural parts of England. The consulate, having become shattered beyond repair, was demolished, whilst the other buildings have been modified after almost every shock of some severity. The first features to disappear were the chimneys, the operation being gradual and extending over several years. Next a roof went and was replaced by one of a light French type, and then followed arches, a tower with a water tank, and a variety of internal decorations. During this period residents in these buildings were in more or less danger, and often sought refuge in their gardens. Now, after an expenditure of many hundreds of pounds, these buildings, the property of our Government, are fairly secure. What is true for this group of structures is generally true for many modern buildings in Japan, and the country is now able to cope with severe disturbances, which, instead of, as formerly, involving outlays of many millions of dollars, may be met by as many thousands, and a proportionate decrease in the loss of life.

SEA WAVES

Many great earthquakes and volcanic eruptions which have originated beneath the ocean have apparently been accompanied by waves, the progress of which around the world has only been arrested by the continents.

In 1868 and 1877 earthquakes causing destruction

along the shores of South America gave rise to sea waves which, 23½ hours later, after a journey of 8,844 miles, reached Japan, where for nearly a whole day, every fifteen or twenty minutes, the sea continued to rise and fall like an unusually high tide. As on these occasions it was impossible to say to what height the next tide might rise, and as there were many traditions of towns having been inundated by a sudden rising of the waters, it is not surprising that many of the inhabitants along the coast sought refuge on the higher ground; and what happened in Japan happened in a greater or smaller degree upon the shores of all countries bordering the Pacific Ocean.

In large bays with a narrow entrance, or in bays and on shores which dip down steeply beneath the water, these movements may be barely perceptible. The localities in which destruction may be expected are naturally those which are nearest to the sea, especially if situated on a gently sloping shore or at the head of a bay which has a wide opening facing the ocean. In such places waves twenty or even eighty feet high may break upon the shore.

The destruction caused by these so-called tidal waves, especially on the coasts of South America and Japan, has often been greater than that resulting from the shaking of the ground.

Those who live on low ground, although they cannot save their property, may often save themselves by taking the advice of the gods of Ise, who, after the disaster which in 1707 overtook Osaka, uttered this oracle : ' At the time of a great earthquake run to a bamboo grove, but at the time of a sea wave seek refuge on a high place.' The people, who had spent days in prayer and in making offerings to the gods, from whom they sought advice, went away disappointed, not because the advice was unsound, but because they knew from experience that the interlacing roots of the bamboo prevented the opening of the ground, and that sea waves did not reach high places.

The *Jishin Nendaiki* or earthquake calendars, a class of publications probably peculiar to Japan, tell us that the eastern coasts of that country have often been ravaged by sea waves, which have carried off from 1,000 to 100,000 people. New editions of these works will contain some account of the waves which on the night of June 15, 1896, inundated the north-west coast of Nipon along a distance of seventy miles, and caused a loss of nearly 30,000 lives. Out at sea these were so long and flat that fishermen did not observe them, but when they put back in the morning they found their villages reduced to heaps of sodden *débris*. At one place four steamers had been carried inland, whilst 176 vessels of various descriptions lined the foot hills. Although we know that some sea waves have been produced by submarine volcanic eruptions, and others possibly by submarine landslips, in this case their origin appears to have been seismic. They came from a district which is well known as the birthplace of many severe shakings, and they were preceded by shocks the movements due to which were recorded in Europe. A whale or a submarine boat submerged beyond a certain depth may move from point to point and not betray its presence by a ripple on the surface, but if the size of the moving mass is at all comparable with the depth of the water this is no longer the case. To explain the waves of 1896, which originated in water reaching to a depth of 4,600 fathoms, all that is required is a sudden displacement of material equal in volume to that which was displaced in Central Japan in 1891.

Now let us ask whether the engineer can avert these disasters. The answer is ' No,' but possibly he may mitigate them. The fact that the only three houses left standing at Kamaishi were storehouses, which relatively to the ordinary Japanese dwellings are substantial structures, suggests the idea that solid masonry may at least palliate a disaster. After the wave of 1854, which partially destroyed the city of Simoda, a sea wall was built further to the south, and this no doubt will protect the town from waves of moderate height. The great point to be attended to when building along coasts subject to inundation of this nature is the choice of a site. If the site is unfavourable a city may have to be removed. After the inundations of 1369 and 1494 such was the fate of Kamakura, a well known village to every sojourner in Japan. Here at one time stood a city boasting of a million people, the palace of a Shôgun, the capital of the empire. Often was it laid waste by fire and sword, but its greatest enemy was the ocean. At the present time Kamakura is a quiet village sheltered by sand dunes and crooked pines, whilst the capital of the empire is Tokyo. All that remains to attest the former magnificence of this pretty hamlet is a gigantic bronze image of Buddha, fifty feet in height, cast more than six hundred years ago, an emblem of solidity, majesty, and peace, the wonder of the engineer, the artist, and the sightseer.

Mary McNeill Scott, writing in the ' Independent,' gives life to Buddha in the following words :

> What do I dream of ? Ah ! the glories gone ;
> Once, all before me, 'twixt the sea and me
> Lay a fair city—rose a Shôgun's home.
> Fair Kamakura, ruled by him and me.
> Jealous the Sea-God ! In one mighty wave
> Swelled his proud heart, the waters rose apace—
> Rose and swept inward ; at my forehead drave,
> Crested the hill tops for a moment's space.
> Only one moment. From the insulted land
> Swift it receded. Ah ! the wreck it bore !
> Oh ! the fair city built upon the sand.
> Oh ! the fair city, seen no more—no more.

CHAPTER X

THE POSITION, CHARACTER, DEPTH, AND DISTRIBUTION
OF EARTHQUAKE ORIGINS

Origins as determined from what is seen or felt—Indications of the
position of origins from overturning, angles of emergence, the form
of isoseismals, and the rotation of bodies—Deductions based upon the
times of arrival of a shock at different stations, and the differences
in time in the arrival of movements of different characters—The
relationship of a meizoseismal area to angles of emergence—The
suggestion of Ōmori—Distribution of earthquake centres in Japan—
Distribution of centres and movement in Tokyo.

EARTHQUAKES which are only felt in an extremely limited
area, and create the impression that a small but sharp
blow had been received by the rocks beneath the observer's
feet, have probably an origin of limited dimensions,
lying at a comparatively shallow depth. Disturbances
which extend over great distances may, on the contrary,
be the result of an effort exerted over a large region,
which may be altogether at a great depth, or have some
part of it at the surface. The Japan earthquake of 1891
was accompanied by the formation of a fault, which on the
surface showed a downthrow of twenty feet and a length
of from forty to sixty miles. The extent of the displacement beneath the surface, the depth to which this
extended, and the total length of the fault are unknown.
The epicentres of the earthquakes plotted in the seismic
survey in Japan are merely approximate centres of areas
where destruction or movement was at a maximum.

In the case of severe shocks, when large bodies have
been overturned or projected, the directions in which these
displacements have taken place have often been plotted as

lines, and the districts in which they intersect been taken as epicentral areas. With earthquakes of moderate intensity, inasmuch as the direction of motion experienced at any point is for many reasons extremely variable, this method fails.

Mallet's well known method of determining not only an epicentre but the depth and position of a centrum by lines drawn parallel to angles of emergence, or at right angles to lines of fracture in masonry structures, might, at least outside an epifocal area, be equally well used for the approximate determination of the maximum slopes of surface waves, and therefore the reliance that can be placed in such a method is not always very great.

The same writer discusses the relationship between the form of the isoseismals and the form and position of the focal cavity. If, for example, we take the latter as being a fault on the face of which some sudden effort is exerted, then the greatest effects upon the surface will lie upon the dip side of the line of strike, and upon this side (' Neapolitan Earthquake,' vol. ii. p. 267) isoseismals will be drawn out to form ovals or distorted ellipses. By reasoning analogous to this, Mr. C. Davison showed that it was probable that the Comrie earthquake of July 12, 1895, originated from a slip of one of the systems of fault lying to the south-east of that place, and trending to the north-west.

Other means of determining the direction from which a shock or shocks have come is to refer to the indication of instruments and the directions in which bodies have been rotated. Directly we attempt to make accurate determinations of the position or depth of an origin from the differences in time at which motion has been accurately recorded at a number of surrounding stations, we encounter a series of problems which are usually more interesting as studies than as aids to exact knowledge.

The most general of these problems is to determine the position of the epicentre, the depth of the centrum, and the velocity of propagation when we have given the time at which a shock arrived at five or more places, whose positions are marked upon a map. The simplest solution to this is by a method of co-ordinates. To describe this and explain the graphical solutions by means of straight lines, circles, hyperbolas, the method of Seebach, the analytical method of Haughton, would be to repeat what has already been published (see Milne's 'Earthquakes,' p. 199).

The inaccuracies which so often accompany the application of these methods are, amongst other causes, due to the difficulties of obtaining a series of time records referring to the same phase of motion, and to the assumption that earthquake motion is propagated with a constant velocity in straight lines from a centrum to points of observation on the surface. When the methods are applied to the sea waves of earthquakes, the velocity of propagation of which is not only comparatively slow, but also more uniform than the elastic and quasi-elastic movements through rocky strata, these objections largely disappear.

If we can satisfy ourselves as to the velocities of propagation of any two classes of earthquake motion—as, for example, the normal or transverse waves, the quasi-elastic waves, the preliminary vibrations, or the sound and sea waves—then by noting the time of arrival of any two of them, we are in possession of factors enabling us to calculate the distance between the point of observation and the origin, and two such distances from points not in a straight line with each other and the origin, lead to a determination of the centrum. Since it is only in extremely exceptional cases that a distinction can be drawn between the first two types of motion (see pp. 91 and 114), the basis of such calculations must rest upon the three remaining types.

An ingenious method of determining the depth of an earthquake centrum suggested by Mallet depends upon the assumption that the meizoseismal zone corresponds to an area in which the emergence of the wave-path at the surface is at the angle where the horizontal component of motion is a maximum.

If intensity varies inversely as the square of the distance from the centrum, the angle of emergence on the boundaries of the meizoseismal zone is 54° 44' 9" ; but if it varies directly as the distance from the origin, this angle becomes 45°. Messrs. Dutton and Hayden, who, from the Charleston Earthquake of 1896, worked with the former of these assumptions, drew many curves of intensity, and showed that the depth of the origin might be taken at about twelve miles. The objection to determinations of this nature are numerous. One assumption is that the disturbance is propagated from a point through an ideal medium. If the earthquake originated along a line, it would probably be more correct to consider the dissipation of energy to the right and left as being inversely as the distance from such an origin. With a radius of surface distribution small compared with the depth of the origin, the loss of energy might be inversely as the cube of radii, measured from such a centrum. The curves of 'intensity' employed by Messrs. Dutton and Hayden may be expressed by

$$y = \frac{k}{h^2 + x^2}$$

where y is the intensity at any point which is at a distance x from the epicentrum and k the depth of the origin. If y, instead of representing energy per unit volume, be considered as being proportional to destructivity due to acceleration, then, as Ōmori has shown, the equation becomes

$$y = \frac{k}{\sqrt{h^2 + x^2}},$$

and with this equation he determined the depths of the following earthquakes :

1. Mino and Owari	.	.	.	1893	7 to 15·6 km.
2. Kumamoto	.	.	.	1889	5·8 to 15·6 km.
3. Ischia	.	.	.	1881	500 m.
4. „	.	.	.	1885	800 m.

From these illustrations it is clear that we are entering
a field of speculative seismology founded upon uncertain
hypotheses. Epifocal areas we can often locate from what
is seen, recorded, and felt upon the surface, but directly
we endeavour to define with any approach to accuracy
the depth of earthquake origins, our deductions, beyond
showing that these are confined to a superficial crust not
more than twenty or thirty miles in thickness, are but
rough approximations (see 'Origins Determined from
'After-shocks,' p. 207 ; Schmidt's 'Methods,' p. 128).

Distribution of Earthquake Origins

Earthquake chronology shows that within the historic
period there is probably no country in the world which
has not been shaken by an earthquake of local origin.

The regions where these occurrences are most frequent
are those in which secular movements are pronounced.
These have been indicated in the chapter explaining the
origin of earthquakes, and all that remains to be done
is to describe the more detailed distribution of seismic
origins as exhibited during the last few years in a country
like Japan, and after that the distribution in an area of
restricted dimensions, such, for example, as the city of
Tokyo.

Distribution of Earthquake Centres in Japan

To determine the number of shocks which are felt per
year in the Japanese empire, which covers an area of
140,000 square miles, I communicated in 1880 with
residents in nearly all the principal towns throughout
the country, asking them to furnish information about the
seismic activity, both past and present, of the districts in
which they resided.

To extend the information thus obtained, bundles of
postcards were sent in the following year to towns and
villages around and to the north of Tokyo, with a request
that every week a card should be returned with a statement of the earthquakes which had been felt.

From the observations thus obtained it was found that for each earthquake a map could be drawn showing the area over which shaking had been sensible, and a close approximation could be made to the position of its origin.

The fact that the greater number of origins were along the seaboard, or beneath the ocean, rather than amongst the mountains and volcanoes forming the backbone of the country, helped to destroy the popular idea of the intimate relationship which was supposed to exist between seismic and volcanic activities.

By 1894 this system of observation had, under Government auspices, been so far extended that it embraced 968 stations, at about forty of which seismographs were established.

The records of these between 1885 and 1892 showed that the empire had been more or less violently shaken 8,331 times, and that Japan could be divided into fifteen seismic districts, in which nearly all these shocks found their origins.

Eleven of these districts lie along the eastern seaboard of the country, or on the face of the steep monocline which sweeps downward beneath the deep Pacific.

Those which contain comparatively few origins lie along the western coast, whilst one follows the line of a steep valley, following the direction of an ancient fault which divides Japan geologically into two halves. Another feature connected with the distribution of earthquakes in this country is that they are most frequent in those districts where there are marked evidences of brady-seismical action, which along the eastern coast is nearly altogether that of elevation. Earthquakes are quite as frequent upon the flat alluvial plains around Tokyo as they are upon the rocky coast and more mountainous districts to the north and south.

What we learn from Japan is but a confirmation of what we learn from the general distribution of seismic activity throughout the world : earthquakes are frequent where we find evidences of secular motion, and where it is not unlikely that such changes may yet be in progress—as, for example, among the younger mountain regions. They may occur along anticlines, as in the Apennines, but they are probably still more frequent along the faces of monoclinal folds, where observation shows that faults are numerous, and beneath which, under the influence of continental load, it is possible that there is an intermittent secular flow of quasi rigid material.

Distribution of Seismic Activity in Tokyo

To determine the extent to which earthquake motion was felt in different parts of Tokyo, I distributed in 1887 through the city and its suburbs, over an area measuring about six miles by five miles, 134 bundles of postcards. Each card, which was addressed to myself, had upon it in English and Japanese the following request: ' If you or your neighbours feel an earthquake, kindly post this card, giving the *date* and the *time* of the shock, and saying whether it was *short*, *long*, a *tremor* or a *jerk* ; were you upstairs or downstairs?' With each bundle, in which there were twenty cards, there was a letter of more detailed instructions. Great care was taken in the distribution of these cards ; and they were all held by persons competent and who expressed a desire to make the necessary observations. Seventy-five observers were situated on high ground and fifty-nine on low ground. The high ground is from fifty feet to a hundred feet above sea level on the western and northern sides of the city, and overlooks the lower part of the city from bluff-like scarps. It consists of thirty or forty feet of loam, thin bands of clay, and sixty or eighty feet of sand and gravel. Below this there is a clay-like tuff rock. The low ground, which is flat almost to sea level, consists of mud, clay, and sand, after which comes tuff. The thickness of these materials lying on the tuff is anything between twenty and 500 feet. In addition to the postcard observers there were many who communicated with me by letter. I also received the records of the

Imperial Meteorological Observatory, from which I could determine the area over which any earthquake extended, and the records from two observatories under the direction of Professor Sekiya. Finally, there were my own observations. Altogether, within the thirty square miles I had about 150 correspondents. The general results of the observations were as follows:

Out of 2,010 postcards which were distributed between November 15, 1887, and May 5, 1888, a period of nearly six months, 103 observers sent in 496 records, 370 of which came from sixty-one observers living on high ground —that is, upon the western and northern side of Tokyo; while 126 records came from forty-two observers living on the low ground.

The average number of records per observer on the high ground was six, while upon the low ground the average was three.

The greatest number of earthquakes was therefore observed by residents on the high ground.

The disturbances which were felt only in Tokyo were at least twenty-five in number. In eight other cases, as the shock was recorded by one observer only, it is possible that a mistake may have been made in observation. All these earthquakes, with the exception of one which is said to have been felt upon the east side of the city, were felt only upon the hilly hard ground upon the western and north-western side of the city.

The disturbances which were felt in Tokyo, and which in addition also shook a large tract of country surrounding the city—in some cases the whole coastline for at least 200 miles—were thirty-six in number.

Now it was extremely curious to find that out of these thirty-six shocks, there were thirty, each of which shook a land area averaging over 10,000 square miles, which were felt only upon the hilly hard ground of Tokyo.

The remaining six, each of which shook an area of about 20,000 square miles, were felt throughout the whole city.

One explanation why so many large shocks should

have passed through the city unnoticed, excepting by seismographs, may be the fact that their average period was 1·85 seconds, while the average period for the six that were felt was ·76 second.

The quick motion was noticed, and the slow motion passed by unrecorded.

A second explanation is that the slower movements reached the surface only where the superincumbent soft materials were thin, while in the thick soft deposits, on the low ground, these motions almost disappeared by absorption.

Out of the total number of earthquakes recorded there were certainly eleven which the Central Observatory failed to record, the reason being, not that the instruments had failed to act, but because the disturbances failed to reach them.

It was clear that there were many small disturbances which had their origin beneath the high ground on the north-western side of the city; and when I lived in that quarter I often felt them as that mysterious little tap described when speaking of the nature of earthquake motion (p. 77).

Since making these observations seven years have passed, and no doubt the mysterious subterranean rappings so often unrecorded have repeatedly announced to the inhabitants of that district that they were living above strata that were intermittently yielding to the effects of strain.

These movements apparently culminated on July 20, 1894, when a severe earthquake of local origin not only did great damage amongst the European brick-built buildings in the lower ground, but caused an equal amount of destruction upon the higher ground.

At the German and English Legations brick buildings were so far shattered that they practically required rebuilding. As pointed out in the chapters upon construction, it is generally known that buildings on soft ground suffer more than buildings on hard ground. The experience of June 20 shows that such a rule requires to be modified for shocks of local origin.

CHAPTER XI

SEISMIC FREQUENCY AND PERIODICITY

Frequency and seismic sensibility—Frequency in Comrie, Kyoto, Tokyo—
The after shocks of 1889, 1891, and 1893—Curves of activity—Frequency in relation to distance from an origin—Meteorological
phenomena—Annual and semi-annual periodicity—The work of
Perrey, Schmidt, Chaplin, Ballore, Mérian, and Mallet—Earthquakes
in relation to the moon and sun—The harmonic analysis of Dr. C.
G. Knott—Dr. Davison's investigations—Dr. Seidl on earthquakes
and barometric gradients—Why our definite information on periodicity is small—The Japan catalogue—Dr. Knott's analysis of the
same—Earthquakes in relation to phases of the moon and tides—
Diurnal and semi-diurnal periods—Periodicity in after-shocks.

Earthquake Frequency or Activity

By the frequency of a phenomena we mean the number of
times it is repeated during a given interval of time. For
example, at Comrie in Scotland in 1844, during the
month of January, twelve earthquakes were recorded, in
the following month there were four shocks, while in
succeeding months the numbers became less.

In the district of Kyoto in Japan during the ninth
century sixty strong disturbances were felt, but since that
time, century after century, such disturbances on the
whole have gradually become less frequent.

High frequency implies a high degree of seismic
sensibility. After a district has reached a certain state of
strain it is liable to yield, which it may do as a series of
small displacements following each other at short intervals,
as, for example, was the case about 1870 in the Yokohama
district. The more general rule is that a district under

strain yields with a sudden crash, which is followed by a number of minor yieldings until seismic sensibility returns to its normal state.

One case may be likened to a bending stick which only crackles, while the other, which is the common occurrence, may be compared to the stick which gives way by snapping, and then crackles as it yields still further.

The case which may compare to the stick which simply breaks and all is over, although not unknown, is comparatively rare. Such a phenomenon apparently occurred at Casamicriola in 1883, in Tokyo on June 20, 1894, and near Sakata in October 1894.

In Japan we have had several excellent examples of the sudden yielding of a rocky mass followed by a long series of after-shocks, indicating that disjointed strata were settling to a state of equilibrium.

On these occasions to the most casual observer it was evident that after the first great blow had been delivered the frequency of the subsequent movements decreased at an increasing rate.

After the terrible disaster of October 28, 1891, no less than 1,132 shocks were recorded during the first ten days. Between the seventieth and eightieth days the number had decreased to eighty-seven, while between the 150th and 160th only thirteen were counted. The decrease, although fairly regular, was not absolutely so ; now and then a shock a little severer than its companions would occur, and would be followed by its own little aftershocks. Since it was clear that the high frequency indicated a high degree of seismic sensibility, it seemed desirable that these after-shocks, which in two years numbered 3,364, should be examined to determine the existence or non-existence of a law governing the decrease in frequency.

This work, which was undertaken by Mr. F. Ōmori ('S. J.' ii. 71), showed results which were more varied and of much greater interest than was at first anticipated.

For purposes of comparison the after-shocks of the Kumamoto (July 23, 1889) and Kagoshima earthquakes (1893)
were also considered.

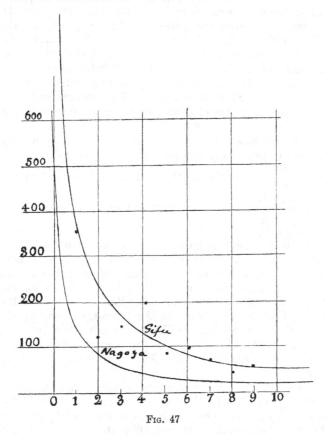

Fig. 47

The total areas sensibly shaken by these three shocks
were respectively 54,000, 6,500, and 5,000 square ri
(1 square ri = 5·7 square miles), which quantities are in
the ratios of 11 : 1·3 : 1.

The after-shocks for these earthquakes for the first thirty days were 1,750, 341, and 279, and for the first two years of the Meno-Owari and Kumamoto earthquakes 3,364 and 834, from which it is clear that the larger the initial disturbance the greater is the frequency of the after-shocks. Fig. 47 shows the after-shock curves for Gifu and Mayorya, which are about 7 and 15 ri (1 ri = 2·4 miles) from the region of greatest disturbance.

The distance measured along x represents intervals of ten days, while the corresponding vertical ordinates indicate the number of shocks which occurred during such intervals.

Mr. Ōmori shows that the curve for the daily activity at Gifu is practically satisfied by the empirical equation $y = \dfrac{519 \cdot 1}{x + 1 \cdot 535}$, while for the monthly activity it becomes $y = \dfrac{16 \cdot 91}{x + 0 \cdot 397}$, where y equals the number of shocks at any time x.

If it is allowed to apply these equations to intervals of time greater than those over which observations have extended, then we have the means of determining the length of time that will be taken before the region will regain the same seismic quiescence that it had before the initial disturbance. Thus for the Mino-Owari earthquake it would seem that there will be one weak shock per day, per week, per fifteen days, or per month, after intervals which are respectively one, four, ten, twenty, and forty years.

How far the higher numbers of this series will prove correct is a matter for observation, but it is not unlikely that a quantity nearer to the truth may be obtained by dividing them by four—that is to say, after about ten years Gifu may experience one weak shock per month, while the same state of quiescence may be reached in Kumamoto in seven or eight years, and at Kagoshima in two or three years.

At the present time in the latter district, after an

interval of about five years, there are about two or three small shocks per month.

The total number of shocks which will occur after a primitive disturbance until a certain stage of activity has been reached is evidently given by the area included between the primitive curve and its ordinates.

For the above three earthquakes, during the intervals required to reach the above-mentioned states of quiescence, the numbers to be recorded will be 4,500, 1,100, and 600. The above investigations, which certainly show a remarkable degree of similarity—at least until the frequency curve has become fairly asymptotic—seem to indicate that on the average the number of 'points of instability' which have to be removed before stability is reached are proportional to the intensity of the initial disturbance, which we know in some instances at least is roughly proportional to the dimensions of the rock fracture at the origin.

Another feature brought out by Ōmori's investigations is the relationship of frequency during the same intervals of time at points which are at different distances from an origin. For the earthquake of 1891 the number of disturbances which occurred at the places mentioned in the following table during the same intervals of time were as follows :

—	Gifu	Nagoya	Tsu	Yoto	Osaka	Tokyo
Number of shocks . .	4,500	2,000	350	140	80	30
Distance in ri from origin (1 ri = 2·4 miles)	7	15	25	25	36	68

From this table it is evident that something may be learnt respecting the frequency of shocks of different intensities, those which travelled to a long distance representing a greater energy than those which travelled to shorter distances.

Also, having given the number of after-shocks, say at points along a coast line, inferences may be made as to the distance of an origin.

Earthquake Frequency in relation to certain
Meteorological Phenomena

E. Knipping, formerly in charge of the Meteorological Bureau in Tokyo, pointed out that the earthquakes observed in that city have been more frequent at or about the times of high wind, an observation which has been confirmed by Dr. Ferd. Seidl (' Mitt. d. Deutsch. Gesell. f. Natur und Völkerkunde Ostasiens,' Band ii. S. 109). Darwin speaks of the earthquakes in certain parts of South America as being regarded as indications of coming rain. In Montserrat earthquakes have followed rains and floods.

Earthquake Periodicity (Annual and Semi-annual)

The analyses of earthquake catalogues with the object of determining whether earthquakes have shown a periodicity in the times of their occurrence has always attracted considerable attention, but notwithstanding the time that has been expended upon these investigations, the positive results attained are very few.

Professor Alexis Perrey, of Dijon, who worked with a catalogue completed some fifty years ago, showed that earthquakes were more frequent at new and full moon (syzgies) than at half moon (quadratures), when she is nearer the earth (perigee) than when she is further off (apogee), and when the moon is on the meridian than when on the horizon.

The difference between the numbers representing the number of times that earthquakes were noted in each of these two periods is, however, not only very small, but there are many instances in which the rules are apparently reversed.

Julius Schmidt found a diminution of earthquakes at full moon, while Chaplin, examining the earthquakes recorded in Tokyo, does not appear to have confirmed any of Perrey's results.

M. de Ballore ('Archives des Sciences physiques et naturelles,' tome xxii. Geneva 1889), who worked with a catalogue of 35,511 shocks, divided the lunar day of 24 hours 50 minutes into eight parts, of which the middle of the first part corresponds to the time of the superior culmination, and found that the number of shocks in any of these parts varied between 5,508 and 5,662.

Lists like these lead us to the conclusion that earthquakes have practically been as frequent during any one of these lunar periods as during any other.

When the occurrence of earthquakes is examined relatively to the position of the sun, which is equivalent to determining their relative frequency at different seasons and months, we find that the greatest number of shocks have been recorded during the winter months, earthquakes being most frequent in the northern hemisphere when they are at a minimum in the southern hemisphere, and *vice versâ*.

The winter law, first pointed out in 1834 by Mérian, implies an annual period. That such a law existed has for long been recognised by those who live in districts where earthquakes are frequent.

Examinations of earthquake statistics made by Perrey and Mallet confirmed the popular opinion, the latter investigator showing that not only was there an annual periodicity, but also that semi-annual times of maxima and minima were marked.

The first writer to subject earthquake catalogues to a more rigid examination than that which follows a classification according to months was Dr. C. G. Knott, who in 1884 pointed out that a classification based upon civil rather than natural divisions of time was arbitrary, and that to obtain the best results the registers should be subjected to harmonic analysis. He gave a simple but sound arithmetical method for separating the annual and semi-annual periods (if such existed), and applied the method to earthquake statistics for a number of countries, with the result of confirming the winter law, and showing

P

that outside the tropics there was also a distinct semi-annual period.

The annual periodicity Knott explains as due to the annual periodicity of long continued stresses over large areas—namely, snow accumulations and barometric gradients. It is also suggested that the semi-annual periodicity may have a connection with the change in barometric gradient. In 1893 Mr. Charles Davison extended Knott's work by applying somewhat similar methods to no less than sixty-two records, forty-five of which belonged to the northern hemisphere, fourteen to the southern, and three to equatorial countries.

In every district, and in all but five records (which are incomplete), the annual period was fairly well marked, the maximum epoch occurring in the winter of both hemispheres.

It is observed that the amplitude of the curves illustrating this period is small for insular districts like Japan and New Zealand, and again for extensive areas in which there may be different districts having the maximum epochs at different dates.

Only three records, which are possibly incomplete, fail to show a semi-annual period, and in fifteen of these cases the amplitude of this semi-annual period exceeds that of the annual period.

It is noticeable that eleven out of these fifteen records include localities like Japan, the Malay Archipelago, New Zealand, West Indies, and the Grecian Archipelago.

If seismic periodicity is considered in relation to intensity, it seems that destructive earthquakes have chiefly occurred during the summer months, while for the slight earthquakes both the annual and semi-annual periods are more marked.

Mr. Davison shows that in ten districts the maximum epoch of seismic and barometric annual periods coincide, in nine districts the latter follows the former by about one month, and in four districts by two months.

In Japan, Tokyo, India, and California the former precedes the latter by about two months.

There are also five other exceptional cases (Scandinavia and Iceland, Great Britain, the district round and including Vesuvius, and Sicily); that is, if we except eight cases out of thirty-one which it has been possible to compare, the general rule is that seismic and barometric maxima coincide. Finally, we meet with the ingenious suggestion that the small amplitude of the annual wave for countries like Japan is accounted for by the fact that many shocks originate beneath the sea, where, because the ocean has time to take up its position of equilibrium as barometric pressure changes, the total pressure on the bottom is fairly constant.

By means of the accompanying table, Dr. Ferd. Seidl[1] shows the marked relationship that exists between earthquake frequency and the state of the barometric gradient in Europe. The earthquakes are those observed between A.D. 306 and A.D. 1842. The gradients are indicated in millimetres per 2,820 kilos. measured from the continent towards the North Atlantic ocean.

	Jan.	Feb.	March	April	May	June
Earthquakes . . .	147·7	138·6	119·4	104·6	94·7	95·4
Gradient · . . .	12·6	8·0	4·2	1·6	−0·2	0·6

	July	Aug.	Sept.	Oct.	Nov.	Dec.
Earthquakes . . .	104·4	101·8	110·2	110·9	123·7	136·4
Gradient . . .	0·4	1·5	5·3	9·2	6·0	9·3

Another point to which the same writer directs attention is the difference in ratio of the number of shocks recorded in different districts at similar seasons. For example, if the earthquakes recorded in summer were denoted by unity, then for Switzerland, Dalmatia, and Italy the winter frequency may be expressed by the numbers 3·3, 1·6, and 1·1.

These differences, especially for neighbouring districts, are certainly remarkable. For the particular cases their explanation may rest in the fact that over the Alps during

[1] 'Die Beziehungen zwischen Erdbeben und atmosphärischen Bewegungen,' *Mitt. des Musealvereines für Krain*. Laibach, 1895.

the winter there is a greater difference in atmospheric pressure, and a greater load by the accumulation of snow, than there is over the neighbouring districts. Another distinction between the Alpine and the two latter districts is the fact that the strike lines of the mountain ranges are in directions far from being parallel to each other, and that therefore they offer unequal degrees of resistance to stresses due to a barometrical gradient, the direction of which may be common to each district. From the little that has been said it appears, therefore, that the only pronounced periodicities which earthquakes present are the annual and semi-annual maxima periodicities.

Considering the labour expended upon the analysis of earthquake catalogues, at first sight it seems strange that the definite results have been so few in number. To explain this, however, we have not far to seek. Although intervals of time approximately following some definite law may be occupied before an area under secular geologic influences may repeatedly reach a state of seismic sensibility, there does not appear to be any valid reason to suppose that in widely separated districts these times should be coincident or necessarily follow the same law. If we, therefore, hope to discover any law bearing upon the recurrence of earthquake susceptibility, it seems necessary that we should first obtain sets of records the entries in each of which refer to the same orogenic fold. The regularly decreasing frequency in after-shocks, dependent on the time taken for disjointed strata to overcome internal friction and the friction on the surfaces of fracture, indicates a law governing the destruction of seismic sensibility, but whether records of the character just suggested can be obtained to throw light upon its creation is problematical. Passing from hypogenic actions as influencing the frequency and periodicity of earthquakes, we will next turn to activities which are epigenic.

Knott has shown how loads due to the piling up of snow and barometrical pressures acting over large areas may possibly explain the winter frequency or annual periodicity of earthquakes, but unless we work with a catalogue extending over very many hundreds of years it is easy to see how this law, if not entirely lost, would at least be only feebly pronounced.

For example, in Japan on the average about five hundred shocks occur annually, but every few years disastrous shakings take place, and in Japan, at least, these have been more numerous in summer than in winter.

After one of these shocks for the next thirty days there may be 1,700 or 2,000 ' after-shocks,' with the result that if we have no means of eliminating such series from the general list for that particular year, a summer rather than a winter frequency may be established for the whole country.

All seismic records up to the present have been of such a nature that investigators of periodicity, although they have classified earthquakes according to intensity, or only considered those above a certain intensity, have not had the necessary materials enabling them to separate shocks which are the immediate result of orogenic causes from those which are brought into evidence by influences exterior to our earth.

Destructive disturbances, their after-shocks, shocks in a district when a frequency curve is practically asymptotic to a time ordinate, and shocks originating in different areas have from necessity been treated *en bloc*.

To show the full value of many influences outside our earth in producing a periodicity, it is necessary, in examining the earthquakes of seismic districts in which frequency is normal, to treat them as distinct from the records of districts where the frequency is abnormal.

For example, in Japan there are at least fifteen districts in which earthquakes originate, and each of these districts is exposed at different times to various external influences. Illustrative of this, we observe that the rise and fall of tide along the coast takes place at different hours. If, therefore, we wish to determine whether earthquake frequency is affected by tidal load, it is clear that each district must be treated separately. For the investigation of periodicity, if such exists in the orogenic factor, we require not simply the number of earthquakes in a district, but numbers proportional to their intensities. The only materials which lend themselves to investigations like those here suggested are the entries in a catalogue [1] of some 9,000 shocks recently drawn up by the author for Japan. For each of these shocks we have the position of its origin, the total area shaken and the time of its occurrence, and from a map which accompanies the catalogue it is seen that nearly the whole of these disturbances have originated in one or other of fifteen districts.

For each or certain of these district groups Knott has formed tables showing the relative frequency of earthquakes throughout the lunar day. The series of numbers thus obtained he has subjected to harmonic analysis. Similar discussions were entered into for tables in which statistics were grouped according to monthly periods, the months considered being the anomalistic, the tropical, the synodic, the sidereal, and the nodical months. The conclusions reached by Knott were, for Japan at least:

1. Earthquake frequency is subject to a periodicity associated with the lunar day.

2. The lunar half-daily period is relatively prominent, and its phase falls regularly in relation to the time of the meridian passage of the moon.

3. There is no certain evidence that the ebb and flow of the tides affect seismic frequency.

4. Therefore we look to tidal stress of the moon as the probable cause of a range in frequency, which, however, does not exceed 6 per cent. of the average frequency.

5. There is evidence of a fortnightly periodicity associated with times of conjunction and opposition of the sun and moon.

---

[1] For the original materials enabling the writer to produce this catalogue he is indebted to Mr. K. Kobayashi, late Director of the Meteorological Department in Tokyo, which controls 968 stations at which earthquakes are recorded.

6. Monthly and fortnightly periodicities appear to be associated with changes in the moon's distance and declination, but from this no definite conclusions can be drawn, because harmonic components fully as prominent exist when statistics are analysed according to the moon's position relative to the ecliptic, with which no particular tidal stresses can be associated.

7. Nevertheless, because the maximum frequency falls near the time of perigee, there is some support to the view that earthquake frequency and the earth distance are closely connected.

Dr. Arthur Schuster compares the amplitudes found by Dr. Knott with those which the theory of probability might lead us to expect, and concludes that if there is an effect coinciding with the lunar day it must be so small that it is hidden by accidental effects, and he arrives at a similar conclusion respecting periodicity in relationship to lunar months ('Nature,' vol. lvi. p. 321, and 'Proc. R. S.' vol. lxi. p. 455).

Knott considers that the relative prominence of the second harmonic amplitudes, especially in regard to the synodic month (conclusion 5 above), is a feature which Schuster's important application of the theory of probabilities hardly does full justice to. One thing is certain, however — namely, the comparative insignificance of periodicities that may possibly be attributable to lunar stresses.

Periods of Short Duration

It is a popular opinion amongst residents in earthquake countries that earthquakes are more frequent during the night than during the day, and if we take a list of earthquakes drawn up from personal observation we find that the surmise is apparently correct.

For example, taking a list of 3,842 shocks, noted in Japan 1885–1890, a few of which were recorded by instruments, we find, if we call the hours between 6 P.M.

and 6 A.M. night, that the ratio of the earthquakes which were noted during the night to those which were noted during the day is as 1·96 : 1.

M. de Ballore, from his catalogue of 37,511 earthquakes, finds the ratio to be as 1 : ·8, but when he divides the shocks into groups according to the Rossi-Forel scale, where No. I. are small disturbances recorded by instruments and No. X. disasters, denoting the night disturbances by unity, he finds that the day disturbances are represented by the following numbers:

I.	II.	III.	IV.	V.	VI.	VII.	VIII.	IX.	X.
1·8	·73	·60	·67	·65	·76	·81	·85	1·27	1·02

From this we see that the night maximum is particularly marked for ordinary disturbances, which is explained on the ordinary assumption that observers are in a better position to note what occurs during the stillness of the night than during the noise and business of the day.

Large earthquakes like IX. and X. are as likely to be noted at both of these times.

The preponderance of day disturbances from example I. is accounted for on the supposition that these have been recorded by delicate instruments which are susceptible to disturbances caused by human movements during the day.

If these records were obtained from ordinary seismographs, such as are used in Japan, to mistake disturbances caused by human agency for those caused by the actual shaking of the ground would seldom, if ever, happen.

From a list of 1,168 earthquakes recorded by seismographs in Tokyo between 1876 and 1891, we gather that 608 of these shocks occurred during the day and 560 during the night—i.e. in the ratio of 1 : 1·08.

If we take only the shocks between 1876 and 1886, the ratio becomes 1 : ·84, that is, its character is reversed, or like the result obtained for shocks recorded without the aid of instruments.

From these examples it appears that the records of

seismographs show earthquakes to be as frequent during the day as during the night—the preponderance in either interval varying in different periods.

If, however, we tabulate the same earthquakes in vertical columns, according to the hours of the day, each column corresponding to a month of the year, an inspection of this table shows that, especially for the winter months, there appears for each twenty-four hours a maximum and a minimum; and passing from month to month, the time of the maximum, commencing at midnight in January, grows later until July, when it reaches midday, while from July to December the time of maximum grows earlier.

Ōmori has shown that there is a marked diurnal periodicity characterising the occurrence of after-shocks, whilst Davison, who subjected twenty-six registers obtained by means of instruments in Japan and the Philippines to harmonic analysis, arrives at the following conclusions:

1. The reality of the diurnal variation of earthquake frequency is shown by the approximate agreement in epoch for the first four components (24, 12, and 8 hours).

2. For ordinary earthquakes there is nearly always a marked diurnal period, having a maximum between 10 A.M. and noon. The semi-diurnal period (with maxima generally occurring between 9 A.M. and noon, and between 9 A.M. and midnight) and other minor components are occasionally important.

3. After-shocks show the diurnal period in a marked manner with a maximum a few hours after midnight.

The four- and eight-hours period for the latter are pronounced. Davison suggests that the diurnal periodicity of ordinary earthquakes may be chiefly due to wind velocity, whilst that of the after-shocks may be due to variation of barometric pressure.

The author is inclined to the opinion that if earthquakes have been frequent at or about the times of high winds, this means that earthquakes may accompany rapid changes in barometric pressure, or are frequent when the district in which they occur is crossed by steep barometric gradients.

Since after-shocks have a four- or five-days period, which roughly corresponds to the larger changes in barometric pressure, and since it is likely that along a fault, possibly 100 miles in length, for many months after its formation there are points at which critical conditions are rapidly being produced, it is not improbable that yielding or accelerated settling should be effected by diurnal barometrical changes.

It must, however, be noted that in Japan these changes, with a range of 2 mm., have their maximum and minimum about 9 A.M. and between 2 and 4 P.M., which are hours far removed from those at which the maximum of the weaker and after-shocks chiefly take place.

Periodicity of After-shocks

From the curves of shocks following the Mino-Owari and Kumamoto earthquakes, Ōmori pointed out that there were periods of maxima of 4·5 days, 33 days, and 6·3 months for the former, and 4·6 days, 33 days and 7·4 months for the latter.

The monthly times of maximum are certainly distinct.

219

CHAPTER XII

SEISMIC PHENOMENA OF A MISCELLANEOUS CHARACTER

Electric phenomena and earthquakes—Appearance of the aurora—Humboldt's observations—Earth currents in telegraph lines—Earthquakes and automatically recorded earth currents—Observations made at the Imperial Observatory, and by the author on atmospheric electricity—Hypotheses as to a possible relation between electrical phenomena and earthquakes—The failure of such hypotheses—The movements of *magnetometers* at the time of earthquakes—Experiments of Ayrton and Perrey. Observations at Parc Saint-Maur—The observed movements are probably due to mechanical causes—The magnetic disturbance following the eruption of Krakatoa—The alteration in isomagnetics observed by Tanakadate—The *sound* phenomena of earthquakes—Suggestions by Knott and Davison—*Emotional* and *moral* effects of earthquakes—Icebergs and seismic action—Changes in the level of lakes or *seiches*.

Earthquakes and Electric Phenomena

Boué, who carefully compared the occurrence of earthquakes with the appearance of auroral phenomena, concludes that there is an agreement not only in their times of frequency but also in their intensities.[1]

Fuchs tells us that during the earthquake of 1808 in Piedmont the air was found to be in a very electric state. Humboldt observed that during the earthquake of Cumana an electroscope quickly showed the presence of electricity in the atmosphere.

The Mississippi and Ohio earthquakes of 1812 are said to have corroborated a belief held in South America that electric discharges in the atmosphere and earthquakes

[1] Boué, ' Parallele des Erdbebens, Nordlichtes und Erdmagnetismus,' in *Sitz der K. A. d. Wissench.*, 1856, iv. 395.

are in inverse proportion to each other. References to luminous appearances in the heavens at or about the time of great earthquakes are very common, as, for example, in Catania 1692, New England 1727, Lisbon 1755, and Naples 1805.

A letter from Mr. Thomas Henry, F.R.S., describing an earthquake felt in Manchester (September 14, 1777) speaks of his wife and others receiving in various parts of their bodies shocks similar to electrical shocks. Subsequent to the shock many people complained of nervous pains and hysteric affections similar to those who have been strongly electrified. Perhaps fright may have contributed to produce some of these effects ('Phil. Trans.' lxviii. 221).

Schmidt says that the maximum frequency of electric phenomena occurs in the middle of October or a few days later, and the minimum about the first week in March. Attention was drawn to the connection between earthquakes and earth currents by Professor W. E. Ayrton, F.R.S., in a communication to the Asiatic Society of Bengal, who observed that the Indian earthquake of 1872 was *preceded* by such strong earth currents on the previous evening in the land lines from Valencia to London that in order to send messages it was necessary to loop the lines. The Egyptian earthquake was also preceded by strong earth currents, and examples of earthquakes accompanying or following earth currents are numerous. By experiment, the writer showed that currents occurring at the time of an earth-shaking might be caused by the mechanical motion creating differences in contact between the earth and an earthplate, the result of which would be varying degrees of chemical action.

The diagrams of automatically recorded earth currents on two lines in Japan, which usually showed a maximum of 1 mil.amp. between noon and 2 P.M., did not show any relationship with earthquakes, the frequency variation being as marked when seismic activity was at a maximum as when it was absent or at a minimum.

An observation of Professor D. Ragona that at the

time of an earthquake there was a current passing through a galvanometer to a rodlike conductor in the atmosphere led the writer to examine the records of atmospheric electricity taken at the Meteorological Observatory in Tokyo. The instrument there used is Mascart's electrometer, charged with 50 Daniells, and connected electrically by a water dropper to the atmosphere. The results showed that whenever Tokyo was near the epicentre of an earthquake there was a sudden displacement of the galvanometer needle, indicating an electro-negative condition in the atmospheric electricity. Not feeling satisfied with this result, I set up a second but similar electrometer at my house. I found that, unless the wire connecting the floating plate with the sulphuric acid was repeatedly washed, the instrument rapidly lost its sensibility, with the result that after a mechanical disturbance it suffered a displacement from which it only very slowly returned.

These experiments were extended by taking for more than a hundred days a continuous photographic record of the difference in potential between water-bearing strata at a depth of about thirty feet and the superincumbent strata, the dry earth resistance being about 15,000 ohms.

At the time of two or three small earthquakes the ' needle ' of the instrument was deflected, but, as in the previous experiments, these displacements found their simplest explanation in the supposition that they were the result of mechanical movement.

Not only do many suppose that earthquakes are accompanied by electrical phenomena, but since the time of Stukeley it has often been suggested that earthquakes are the immediate results of electrical discharges. A special pleader for those who seek to explain what they do not understand by an appeal to such phenomena, might call attention to the cubic miles of our atmosphere which are shaken at the time of thunderstorms, and suggest several hypotheses which are not altogether devoid of facts for their support.

One supposition is that a stratum of cloud raised to a high potential might by inductive stress result in the collapse of a rocky crust beneath, which at the time was in unstable equilibrium. Inasmuch as the maximum pull exerted between the two could not exceed that which would bring the clouds downwards, it follows that the forces involved are too insignificant for serious consideration.

A second hypothesis is that beneath volcanic regions watery steam may be escaping through fissures from regions of high pressure to those of lower pressure separated by insulating material. By such hydro-electric action electricity would be developed, and if we add to this condenser inductive action, conditions culminating in violent discharges might be reached.

In support of such an hypothesis it might be pointed out that certain hot springs at volcanic foci *apparently* show a potential markedly higher than that of the surrounding ground. Earthquakes are *sometimes* less frequent during the wet seasons, when it may be supposed that distant areas are reduced to the same potential. As an attempt to push the defence of this popular theory still further, we might add the fact that since California has been covered with railroads earthquakes have been less frequent. The extravagant hypothesis that metal rails have equalised potentials in distant districts does not, however, find its support in Japan.

Although it would be difficult to deny the probability of the existence of difference in potential between points deep underground and between such points and the surface, the facts that can be adduced to support the view that earthquakes result from the equalisation of electrical potential are few in number and weak in character. Moreover, varied and numerous phenomena which we should expect to accompany such disturbances have never been observed, so that, if we except the production of minor phenomena like the luminous appearance accompanying the friction of moving masses of rock, any hypothesis connecting earthquakes and electricity appears to be quite untenable.

Earthquakes and Magnetic Disturbances

About two hours before the destructive Japan earthquake in 1855 the owner of a spectacle shop in Tokyo (then Yedo) observed that a magnet dropped some pieces of iron which had been attached to it, an observation which led to the construction of a magnetic seismoscope.

After the earthquake of Cumana, 1799, Humboldt observed that the dip was changed forty-eight minutes, and Mallet refers to similar observations made at Lima.

In 1822 Arago and Biot observed movements in magnetometers at Paris which coincided with slight shocks in Switzerland. In Italy, Sarti, Malvasia, Rossi, and other observers all testify to remarkable magnetic phenomena taking place at the time of earthquakes.

Mario Baratta, writing in March 1889,[1] after a long and patient examination of a catalogue of occasions on which electrical and magnetic phenomena have preceded, accompanied, or followed earthquakes, concludes that the loss in portative power of magnets is not produced by mechanical movements, but is an effect of some extraordinary earth current produced at the time of the earthquake. Special experiments carried out by Professors Ayrton and Perry many years ago in Japan to determine whether changes of this order could be observed in magnets at the time of an earthquake gave a negative result.

Although a copper bar having the same form and a similar suspension to that of the magnet in a magnetograph at Parc Saint-Maur was not disturbed at the time of three earthquakes whilst the magnet was disturbed, it need not necessarily follow that the shaking was accompanied by magnetic effects. As pointed out by G. Agamennone, the moments of the two bars were different, and a back and forth motion might occur with

---

[1] *Società Geologica Italiana, Bollettino*, vol. ix. fasc. i.

a period that would cause movement in one bar but not in the other.

With the Assam earthquake of July 12, 1897, at Bombay, which was well outside the area of perceptible shaking, the declination, horizontal force, and vertical force magnetometers were greatly disturbed. The director of the observatory there, Mr. N. A. F. Moos, who carefully discusses these records, concludes that they are chiefly the result of magnetic rather than mechanical action. Similar disturbances were noted in Batavia.

The examination of the photographic records of magnetographs at the Tokyo Observatory only show changes that may be explained on the assumption of mechanical disturbance. At this observatory about fifty or eighty earthquakes are felt and recorded every year. In the magnetograms for declination, and occasionally for horizontal force, irregularities accompany a certain number of these movements, but it is seldom that the records for dip are disturbed. The Riviera earthquake of 1887 is recorded as having *simultaneously* affected the magneto-graphs at a number of magnetical observatories in Europe. It must here be remembered that very small intervals in time are not easily measured on an ordinary magnetic record, and therefore it is possible that the disturbances may only *appear* to have been simultaneous. The magnetic disturbances following the eruption at Krakatoa in 1883 travelled eastwards at rates varying between 761 and 739 miles per hour, suggestive of that at which a sound wave might be propagated through the air, but far removed from the rates at which elastic vibrations are transmitted to great distances. Dr. Charles Chree tells me that the character of this disturbance as shown on the magnetographs at Kew was that of a rounded hump of no rapid curvature, indicating an effect extending over a considerable time. Professor John Perry observes that the rate of transmission approaches that of the equatorial velocity of the earth, and therefore may indicate an induc-tive effect from the outside. At Batavia, near to the origin, the magnetic effects were large and irregular, and might well be attributed to the fall of magnetic ash, and if we follow out the suggestion of Professor Perry and remember the dust cloud which seems after the Krakatoa eruption to have enveloped the world, it is not improbable that the magnetic perturbations at other observatories may find a similar explanation.

A curious fact connected with the disturbances observed in the records of magnetometers is that the instruments at one observatory are often disturbed, whilst similar instruments at other stations do not show corresponding effects. For example, the instruments at Utrecht are fairly often disturbed at or *before* the time of unfelt movements of earthquakes which possibly had their origin in Japan, whilst corresponding instruments at Greenwich, Stony-hurst, and other places have remained quiescent.

Assuming similarity in the instruments and the adjustments of the same at each of these stations, and assuming also that the character of the wave motion at each is very similar, we might suggest as a hypothesis that the materials which are disturbed around Kew are generally more magnetic than those which are disturbed around the other stations. Should this be true for Utrecht, it might also be true at Wilhelmshaven and Potsdam, where magnetic disturbances frequently accompany earthquake motion. That the re-arrangement of large masses of the earth's crust may be accompanied by certain slight changes in the isomagnetics of a district is possibly shown by the alterations in horizontal force observed in the Owari district (Japan) by Professor Tanakadate after the earthquake of 1891.

In connection with this subject we must remember that earthquakes represent the relief of enormous stresses within the crust of the earth, so that it is not altogether improbable that alterations in the strains of rock masses may be accompanied by magnetic changes similar to those exhibited under like conditions in magnetised iron or nickel.

Mr. K. Nakamura, Director of the Meteorological

Q

Observatory in Tokyo, points out that the magnetographs at the stations Sendai and Nagoya exhibited unusual disturbances before the earthquake and great sea wave of June 15, 1896. These attained a maximum about nineteen hours before the earthquake, which took place at about 8 P.M.

Similar effects were observed with a maximum thirty-three hours before the earthquake which occurred on August 31, 1896, at 5 P.M.

These disturbances were most marked at the station nearest to the earthquake origins, and they were therefore, as might have been expected, effects which were fairly local. The fact that the maximum occurred some time before the general relief of strain is hardly what would be anticipated. Such changes have not been observed with or before small earthquakes.

Not only have magnetic needles been disturbed before an earthquake and at the time when earth waves have been slowly moving the area on which they are situated, but there are reasons for suspecting that there may be subterranean operations at work which affect the secular variation of the magnetic elements. As a possible illustration of this Captain E. W. Creak, F.R.S., gives for the period 1880–1885—when there were heavy earthquakes in Japan and Manila, and the gigantic explosion of Krakatoa took place—the following notes respecting observations at the three following places, in each case the north seeking end of the needle alone considered.

Bombay.—Until 1883–85 the needle was moving eastwards. It then stopped, since which it has been moving westwards at an increasing rate. In 1881 there was a sudden change in the dip, the needle going down at an increased rate.

Hong Kong.—Until 1875 the needle was moving eastwards. Then there was a rest until 1880, when it began to turn westwards. The dip needle moved downwards until about 1880, since when it has turned upwards.

Batavia.—Until 1884 the needle was moving eastwards, when it became stationary. It is now moving westwards. The dip needle was moving moderately upwards until 1881, but it has now greatly increased.

It is certainly difficult to imagine that adjustments in a magnetic magma beneath the eastern coast of Asia should be accompanied by magnetic disturbances at a place so far distant as Bombay, and therefore, as Captain Creak remarks, the coincidences here recorded may only be accidental. Nevertheless, we know that many lavas are highly magnetic and that the isomagnetics in the vicinity of an active volcano like Ganjusan in north-east Japan have changed at an abnormally high rate, so that to seek for a connection between certain local magnetic disturbances and the mechanical displacements or the physical or chemical changes in a subjacent magnetic material is apparently a legitimate investigation.

Volcanic activity shows that there are changes in the arrangement of magnetic materials, whilst earthquakes which cause the world to palpitate for several hours, and agitate the surface of an ocean like the Pacific for a period of two days, indicate sudden suboceanic displacements of materials the enormous volume of which it is difficult to estimate.

Sound Phenomena.—Nearly all large earthquakes have been preceded, accompanied, or followed by sounds, the nature of which has varied with the position of the observer and the locality in which he has been situated. They have been described as being like thunder, the rattle of musketry, the rumbling of a waggon, the escape of steam, &c.

These sounds are more common in mountain districts than on the plains, and I formerly attributed them to minute quickly recurring vibrations which precede heavy disturbances, and which, in all probability, are continuous with the preliminary vibrations recorded by seismographs. They appear to be transmitted from their origin to the observer, not through the atmosphere, but through the rocky crust of the earth.

After the Gifu earthquake, at a distance of from ten to twenty miles from its origin, I heard booming sounds every few minutes. These preceded small shocks by intervals of one or two seconds. Sometimes the shock and the sound were simultaneous, and often there were

sounds without shocks. Dr. C. G. Knott, who has closely
examined the theory of elastic vibrations in relationship to
the production of sound, traces sound phenomena to rapid
vibrations of the ground, so rapid as to be inappreciable
on our seismographs. These vibrations, whether compressional or distortional, are transmitted to the air as
compressional vibrations, the waves being refracted nearly
vertically upwards, whatever be their direction of incidence,
because of the much smaller speed in air than in rock.
This explains why the sound always seems to come from
below. Dr. Charles Davison traces the origin of sound-
producing vibrations to the lateral margins of the fault
area, where the initial amplitude of vibrations depends
upon the amount of slip. These amplitudes being small,
the waves will necessarily be short in period. I would
suggest that the mechanics of production may be similar
to that which produces sound when we rub our finger on
the edge of a finger glass. In this way two surfaces of
rock may slowly rub across each other, setting up elastic
vibrations, but not necessarily producing any sensible
movement on the surface of the earth.

Although the mysteries of haunted houses have, after
long searching, been traced to the flapping of the halliyards
on a flagstaff, or the intermittent gurgling of a spring
beneath the basement, I am aware that in at least two
instances ghostly sounds have been identified with seismic
sounds, which in country districts might from time to
time be heard at one house, whilst at another a mile
distant nothing might be noticed.

Emotional and Moral Effects of Earthquakes

A newcomer to an earthquake country usually expresses
surprise that the small disturbances he experiences should
create alarm, but after frequent repetitions with, sooner
or later, experiences of shakings that are moderately
severe, the feeling of indifference, if not replaced by one
of alarm, at least gives way to one of anxious inquiry.

When a slight vibration commences we do not know whether it will die out as it has begun, or whether it will culminate in a shock of unknown magnitude, so that the smallest tremors will often cause anxiety about what may happen next.

A disastrous shock will throw the weaker members of a community into a state of terror or hysterics, and at every little shock, perhaps, for the remainder of their lives they will either be so far unnerved that they do not move, or else, seized with alarm, they will seek a place of safety. I am acquainted with two cases which, in consequence of the nervous excitement produced by comparatively small disturbances, terminated fatally.

Buckle and other writers have discussed the effects produced by displays of volcanic and seismic activity upon the mental and moral character of nations cradled amongst these natural terrorisms. An appeal to statistics shows us that calamities which have occurred in earthquake countries have, in all likelihood, been sufficient to shake ideas respecting permanency and to create feelings of carelessness for the morrow. In Japan alone, on October 28, 1891, in thirty seconds the country lost from thirty to fifty million dollars, 9,960 people were killed, and the wounded numbered 19,994; 128,750 houses, without counting temples, factories, and other buildings, were levelled with the plain, landslips stripped the mountains of their forests, valleys were compressed, lakes were formed, the strongest engineering structures gave way, and the country was left fractured, fissured, and tossed into a sea of waves. Between 1783 and 1857 the kingdom of Naples lost at least 111,000 of its inhabitants, and Mallet estimated in 1850 that during the preceding 4,000 years thirteen millions of people had been swallowed up or killed by earthquakes.

On June 15, 1896, Japan lost some 29,000 of its people by sea waves produced by submarine seismic activity.

We see in the Japanese myths explaining the origin of

earthquakes, and in the drawings of the fantastic monsters
which create these disturbances, direct effects upon the
imagination. In the festivals at temples to commemorate
these great disasters, in the prayers which have been formulated, in the repeal of taxes to appease an angry deity, in
the sermons which have been preached, and in the form of
mountain worship, we see how religion and morality have
been influenced by seismic and volcanic phenomena.

When a European community is overtaken by a
disaster like that which visited Japan in 1891, women
become hysterical, men lose their nerves and behave as
if they had lost their reason. On the authority of
Dr. Julius Scriba, we learn that amongst the Japanese
on this occasion, the result of nervous excitement
showed itself in the form of tetanus, spinal and other
troubles, rather than in any general mental paralysis. A
Japanese is not nervous before a public audience or the
surgeon's knife, nor does he show unreasonable excitement
at the time of a great earthquake. If from time to time
in England or in any other European country cities were
levelled, coasts were inundated, mountains flowed like
water, whilst everything which we regard as permanent was
repeatedly reduced to ruin, it would seem natural that ideas
of permanency would be destroyed, a carelessness for the
future might be engendered, and a timidity might be
established amongst the weaker members of a community
which would handicap them in the struggle for existence.
The general temperament of a nation is no doubt largely due
to its environment, and it is not unreasonable to suppose
that serenity of demeanour and carelessness of the future
may hold some relationship to repeated exhibitions of
seismic and volcanic energy.

Icebergs and Seismic Action

In a paper read before the Royal Society of New
South Wales, September 4, 1895, Mr. H. C. Russell shows
that in particular years—as, for example, in 1854 and
1891—not only has there been a remarkable increase in

the number of icebergs recorded by vessels in the Southern
Ocean, but some of them have been of gigantic dimensions,
one, for example, measuring sixty by forty miles. The
fact that these large bergs have had similar triangular
forms suggests the idea that they may have been cast in
the same mould. After discussing by the usually accepted methods, the rate at which bergs may be formed,
their dimensions, and the time taken in their dissipation,
Mr. Russell concludes that the appearances noted in these
particular years cannot be accounted for by any of the
usually accepted explanations, and suggests that the
immense masses may have been broken off at irregular intervals corresponding in time to those of unusual displays
of seismic activity in the Antarctic continent. Another
explanation is to suppose that ice and snow accumulate
upon slopes until critical conditions are reached, when the
whole slides down like snow slides off a roof.

Rapid Changes in the Level of Lakes

That the levels of lakes are from time to time subject to
rapid oscillations was observed in 1830 by Duillier. From
the fact that by the retreat of the waters the shores became
dry, these movements were called *seiches* in Switzerland.

In 1804 Vaucher arrived at the following conclusions
respecting these movements. Seiches are more or less
marked in all lakes; they occur at all seasons and hours,
but particularly in the spring and autumn. The greatest
oscillations are, however, in July, August, and early in
September. The governing cause is the condition of the
atmosphere. Their duration is variable, and their amplitude
varies considerably even on the same lake.

At the present time Mr. Napier Denison, of Toronto,
is studying the movements of Lake Ontario, which he
shows may be regarded as a sensitive barometer ('Proceedings of the Canadian Institute,' January 16 and
February 6, 1897).

For one of the most systematic studies of seiches we
are indebted to Dr. F. A. Forel, who for many years

recorded by means of tide gauges the rapid alterations
so often observable on the lakes of Switzerland. Speaking
of the Lake of Geneva this observer tells us that, independently of the waves produced by wind, the surface
of the lake is never completely at rest.

One class of movements have periods between a half
and four minutes, and fall between the ordinary surface
waves and the longer period motions. These latter,
which are the most pronounced, are of two types—the
transverse and longitudinal.

The transverse seiches may be simple and regular, with
periods of ten minutes ; but sometimes they are irregular,
varying in amplitude, and with periods as small even as
two minutes. These irregular movements find an explanation in the interference of two or more simple seiches.
The amplitude of a simple seiche varies between 0 and
124 mm., and its duration is six or eight hours. It
commences suddenly, and shows a maximum in the first
oscillation. History records seiches in Geneva which
have reached heights of from 1 to 1·9 metres.

Longitudinal seiches are comparatively rare ; they
have periods of seventy-three minutes—at Geneva amplitudes of 10 or 20 cm.—and continue for two or even
four days. If l is the length or breadth of a lake, h its
average depth as measured along this length or breadth,
and t the time in seconds of a semi-oscillation of its
water in the same direction, these quantities are connected
by the well-known formula

$$t = \frac{l}{\sqrt{gh}},$$

or

$$h = \frac{l^2}{gt^2}.[1]$$

The fact that the average depth of a lake, as determined
from the observed time of the oscillations of its waters,

[1] 'La Formule des Seiches,' par M. le Dr. F. A. Forel, *Archives des
Sci. physiques et naturelles*, t. xiv. p. 203.

closely accords with the directly measured value indicates that seiches are natural periodic pendulum-like movements in lake waters.

Seiches are more marked during the winter than during the summer, and when the barometer is low than when it is high. Rapid changes in local barometrical pressure may be sufficient to produce the smaller seiches, whilst the larger ones are, as pointed out by M. Ch. Dufour, the results of intermittent falls in pressure. Winds influence the production of seiches, especially those with a marked vertical component which may accompany the commencement of storms.

In addition to meteorological causes, certain seiches find an explanation in movements of the lake basins by local or distant earthquakes, avalanches, and landslips.

REFERENCES

Seismic, Magnetic, and Electric Phenomena, J. Milne, ' Seis. Journal,' vol. iii. p. 23.

Earthquakes in Connection with Electric and Magnetic Phenomena, by J. Milne, ' Trans. Seis. Soc.,' vol. xv. p. 135.

' British Association Reports on the Earthquake and Volcanic Phenomena of Japan,' by J. Milne, 1890, 1891, 1898. In this last Report the records obtained from a number of magnetic observations are given in detail.

The Volcanoes of Japan, J. Milne, ' Trans. Seis. Soc.,' vol. ix. pt. 2, p. 178.

' Nature,' Jan. 20, 1898, vol. lvii. p. 273.

CHAPTER XIII

SLOW CHANGES IN THE VERTICAL

Changes in the vertical noted at astronomical observatories—Greenwich, Cambridge, Neuchatel, Berne, Sydney—Annual periodicity of these changes—Observations of d'Abbadie—Tidal effects computed by G. H. Darwin—Observations of Plantamour—The diurnal and annual changes—Observations at Berlin and in Japan—The water level at the geodetic institute at Potsdam—Changes observed by von Rebeur-Paschwitz at Teneriffe, Potsdam, Wilhelmshaven, Strassburg —Annual change at Nicolaiew—The author's observations in Japan, made in caves, in alluvium underground, and on the surface—Relationship of these changes to geological structure, fluctuations in temperature, underground water, evaporation and condensation of moisture, and to barometrical pressure—The creep of earth to lower levels.

FOR many years past astronomers have had forced upon their attention the fact that well constructed piers, on which transit and other instruments are placed, show slow changes in position corresponding to alterations in the level of their upper surfaces and deviations in azimuth.

In 1782 S. J. de Silvabelle directed attention to the apparent variations in the position of objects, as seen at different times through a telescope, which in 1813 was followed by Conte Moscati's account of annual and diurnal changes in level. He notes that the latter are only pronounced in clear weather, and, with certain observers who followed him, is inclined to the opinion that these changes may be influenced by fluctuations in underground water.

In 1848, Henry, who examined the changes in level of transits between the years 1833 and 1842 at Cambridge and 1836 and 1845 at Greenwich, concludes that for both places the west γ of their instruments are about 2·5 seconds higher at the vernal than at the autumnal equinox. Also at the former epoch an azimuthal deviation of about two seconds towards the south is attained.

At Greenwich Sir George Airey established systematic observations for the determination of these slow movements. An account of the variations in azimuth and level of the meridian circle is given by Ellis, who shows that although the changes recorded do not correspond in period to the thermometrical changes in the atmosphere, they closely follow changes in temperature observed at a depth of about twenty-five feet, where the maximum occurs in November and the minimum in June. But, as M. d'Abbadie remarks, we do not see why the coincidence should be marked at this particular depth. M. Challis observes that the movements are apparently less with piers founded upon sand than they are when the foundation is on clay or rock ('Lectures on Practical Astronomy,' p. 21, London 1848).

From observations made with two neighbouring transits, Ellis shows that the changes in azimuth are partly due to the instruments, and that the months of maximum and minimum do not closely correspond.

At Neuchatel M. Hirsch observed an annual oscillation with a transit instrument of twenty-three seconds, and a change in azimuth of seventy-five seconds, and somewhat similar changes have been observed at Berne.

Observations made by H. C. Russell, F.R.S., at the observatory in Sydney, which stands on a hill of sandstone rock, show that the level of the transit has a regular annual variation, the eastern side rising in June to its maximum, which is about ten seconds. This movement corresponds in its direction and time of occurrence to what has been observed at Greenwich and *some* other European observatories. The change in azimuth does not show any annual periodicity, but increases gradually at a rate of three seconds to five seconds per year. In this respect the Sydney observations differ from those at Greenwich and other observatories.

For various reasons Mr. Russell is assured that the observed changes are not due to the heating of the rock on which the observatory is built. Since the movement at Sydney corresponds in time to that observed at Greenwich and other places in Europe, it follows that similar changes are taking place during the winter of one hemisphere and during the summer of the other, and that the cause of these annual movements is, in all probability, not to be sought for in changes of temperature.

It was observations such as are here described that led M. d'Abbadie, Plantamour, whose methods of observation have been described pp. 45 and 46, and other observers, to make the recording of changes in level a subject for special study. The work of M. d'Abbadie is of special interest. After allowing the masonry cone above the basin of mercury five years to settle, he commenced operations in 1868. Although both the atmosphere and the neighbouring ocean, distant 400 metres from the observatory, might be calm, it was seldom that there was perfect tranquillity of the surface of the mercury. Sometimes sudden *frétillements* or jumps were observed in the position of the reflected images. Out of 359 comparisons, 243 indicated that a change of level was probably due to the attractive influence of the tide, which when it rose 2·9 metres caused a deflection of about 0·18 second.

There are, however, fifty-seven cases in which the gravitational effect of the tide seemed to produce repulsion of the mercury, from which it may be inferred that the tidal effects are frequently eclipsed by greater effects which occur simultaneously. On one occasion a change of 24 seconds was observed to take place in a period of six hours. Between January 30 and March 26, 1872, the change reached 4·5 seconds, and these changes could not be traced to astronomical or thermometrical causes.

Professor G. H. Darwin, in a report to the British Association, 1882, computes the depression of the surface and slope of an area like the Atlantic, by the rising of the tide along a shore like that of Europe. The breadth of the ocean is taken at 3,900 miles, the tide at 40 cm., and the rigidity of the yielding coast as being greater than that of glass. Professor Darwin shows that since the true deflection of a plumb line due to slope would be augmented by the attraction of the water, the amplitude of the oscillations observed would be one and a quarter times those given in the following table, and between high water and low water the changes would be two and a half times these quantities.

Distance from mean water mark	Slope
10 m.	0″·0504
100 m.	·0403
1 kilom.	·0302
10 „	·0202
20 „	·0170
50 „	·0131
100 „	·0101

The deflections observed will increase proportionally to the height of the tide and decrease with the width of the sea. The land regions remain nearly flat, there is a sharp change in curvature along the shore line, and the slope is greater beneath mid-ocean than on the land.

M. Philippe Plantamour commenced his observations at Sécheron, near Geneva, where a level was placed in an east-west direction upon a concrete floor.

Subsequently two levels were placed in parallel positions on a *chevalet de fer* in a cellar at M. Plantamour's house, where the changes in temperature were exceedingly small.

In the 'Philosophical Magazine' for February 1889 Dr. Charles Davison summarises and discusses these observations. Although observations were commenced in 1868, the years considered in this paper are from 1878 to 1886 inclusive.

During the first two years the levels were read five times a day, from which the character of the diurnal oscillations were determined. After this readings were only taken twice per day, the hours chosen being those at about which the maximum and minimum excursions were reached.

In addition to the discovery of a diurnal change in level, which will be discussed in another chapter, other important results were obtained. These were as follows.

First, an annual periodicity in the tilting of levels was recorded. Between January and April the eastern end of a level reaches its lowest point, after which it rises until some date between July and October. In 1879 the total annual amplitude for the east-west levels was $28''{\cdot}08$, and for the north-south levels $4''{\cdot}89$.

Secondly, it was observed that the greatest change of inclination during a year approximately coincided with the direction of the average slope of the ground on which the observing station was situated.

Thirdly, it was seen that year after year the bubbles of the levels did not return to their starting points, but were gradually shifted towards the north and east.

A rise or fall of external temperature was followed at an interval of from one to four days by a rise or fall of the eastern end of east-west level, and what was true for short period excursions was generally true for those which were seasonal.

The movements of the north-south level, which are comparatively small for the annual excursions, follow a law similar to that shown by the east-west levels, the south end rising in summer and falling in winter. For changes in temperature of short duration this movement is, however, reversed.

After a critical examination of the effects likely to result from changes in temperature, Mr. Davison concludes with M. Plantamour that it is extremely probable that they are sufficient to produce the periodical movements which have been observed.

The movement which year after year is steadily taking place in one direction, and which hitherto has not shown any tendency to be periodic in its character, may be local or widespread.

If it prove to be local, we have in M. Plantamour's observations measurements of the ' infinitesimal changes which culminate in a great mountain chain.' In connection with this it may, however, be mentioned that, as the result of observations extending over forty-two years at the Berlin Observatory, Foerster concludes that an eleven years' period may exist in these slow movements.

In my British Association Report of 1885 I refer to observations which were made upon two sets of astronomical levels installed at right angles to each other in Japan.

One set of these records, which extended over two or three years, is in the possession of the Meteorological Department in Tokyo, and may yet be subjected to analysis. The other set was lost by fire.

Inasmuch as these records have never been closely examined, it cannot be said that they either confirm or disprove the results obtained by M. Plantamour. The only results of importance derived from these observations, which were discontinued for reasons stated on p. 46, were as follows:

1. The bubbles showed considerable changes in position, and certain of these changes preceded earthquakes. After an earthquake the position of the bubble of a level was often changed.

2. The greatest motions were obtained during the coldest part of the year, which is the season of earthquakes, and during which the barometric gradient between Siberia and the Pacific Ocean is the steepest.

3. The bubble of a level continues to move long after the sensible motion of an earthquake has ceased.

4. When the barometer is very low—as, for instance, during a typhoon—the bubble of a level may be distinctly seen to pulsate back and forth through a range of ·5 mm.

Up to the present the water level established at Potsdam by Dr. Kühnen has exhibited an annual periodicity with a continual northerly rising of about 10 mm., but as the director, Dr. Helmert, writes me, the observations

have not yet been continued for a sufficient length of time
to determine whether the recorded changes are in reality
due to a general alteration in the earth's surface.

The late Dr. E. von Rebeur-Paschwitz showed that
his horizontal pendulums, in addition to showing a daily
oscillation, showed changes in their zero points. By
comparing curves showing the position of the zero points
with those of temperature and barometric pressure, it was
found that at Wilhelmshaven a change of 1 mm. in the
latter corresponded to a change in the vertical of $0''{\cdot}29$,
and these changes occur simultaneously.

The effect produced by a change of 1° C. in temperature
is exactly double that produced by the barometric change,
so that the latter is often either masked or intensified by
the former.

At Teneriffe (Puerto Orotava) both these meteorological
effects are small, each being represented by changes of
$0''{\cdot}03$. whilst at Potsdam the effect of temperature only is
visible.

The Teneriffe observations may possibly indicate a
yielding of the mass of the volcanic cone on the flanks of
which the observing station was situated, the slope to the
mountain increasing as the external pressure diminishes.

In addition to these changes in the vertical, which
from the nature of their apparent causes are completed at
irregular intervals not likely to exceed a few days, there
are other displacements to complete each of which may
occupy several months. Thus at Potsdam, during April
and part of May, the column supporting an instrument
moved towards the west, after which it turned and completed a tilt of $11''{\cdot}2$ towards the east.

At Strassburg a von Rebeur pendulum in charge of
Professor Becker, director of the observatory at that place,
showed for a period of nineteen months a curve of wandering similar to that for a curve of temperature; but, since
the minimum of temperature is reached from one and a
half to two months *before* the minimum in the curve
showing the displacement of the pendulum, whilst its

maximum is reached about four months *later*, the relationship between the two becomes obscure. The pendulum curve closely agrees with one deduced from observations with a level, but it is widely different from one showing the changes in the Nadir. Between April 1892 and April 1894, the pendulum, although showing one strong northerly motion, was displaced 143″ towards the south.

From these and other observations, von Rebeur-Paschwitz is inclined to think that the general form of the true oscillation of the plumb line is approximately represented by an ellipse the long axis of which lies between E.W. and N.W.-S.E.

The annual oscillation observed at Nicolaiew does not exceed three or four seconds, and Professor Kortazzi is of opinion that it may be explained by the inclination of the upper layers of the ground accompanying annual changes in contraction at the base of the pillar, which is founded fifteen feet below the surface. In spring and summer the movement is southwards and in winter northwards. The annual change in temperature in the cellar where the pillar is founded is 13·5° F.

The last set of observations bearing upon changes in the vertical having periods greater than twenty-four hours is that made by me in Japan.

The primary object of these observations — which indicated wanderings in the zero point of pendulums, diurnal waves, and earthquakes which had originated at great distances—was to record earth tremors.

With von Rebeur-Paschwitz on the other hand, the idea was different, his horizontal pendulum being set up originally for the purpose of measuring deflection due to the gravitating influence of the moon.

The best results were obtained from horizontal pendulums, the installation of which is more fully described under the section relating to diurnal tilting.

At Kamakura two pendulums were set up in a cave, excavated in fairly hard tuff rock made up of a series of comformable beds dipping 30° N.E.

R

One pendulum was placed to record motion parallel to this direction and the other at right angles to it. The former, although usually the least sensitive of the two, not only recorded the greatest amount of earthquake motion, but showed the greatest wandering of its zero point.

The daily temperature variation in the cave was about 1·5° C. The movements observed had usually periods of from forty-eight to seventy hours. One which occurred between February 28 and March 3 indicated that the dip of the rocks had increased and then decreased through an angle of 4″·08. The movement at right angles to this was 2″·88.

These displacements could in no way be connected with variations in temperature in the cave, with sunshine or want of sunshine on the outside, or with rainfall.

The most important feature in the records is the fact that the greatest movement was recorded in the direction parallel to which it may be supposed yielding would most readily take place. The fact that local earthquakes were frequent when the movements were great suggests a line of investigation deserving the closest attention of those who may endeavour to predict such phenomena.

A comparison of the wanderings between January and May 1895 of two pendulums both installed on the alluvium in Tokyo, but one in an underground chamber where the daily change in temperature was slight, showed that the displacements of the zero point, taken at intervals of from three to five days, of the underground pendulum was usually greater than the other. When the movements exceeded 1 second, they nearly always agreed in their direction, although the distance between the observing stations was about 1,000 feet. When the movements were less than this the cases of agreement and disagreement were nearly equal.

The total movement of the instrument on the surface corresponded to a lifting of the ground on the north-east side through an angle of 6″·76, whilst the corresponding motion underground was 15″·94.

Since the movements underground, where changes in temperature were small, exceeded those recorded on the surface where changes in temperature were large, and since it was found that the deflection of a pendulum caused by artificially raising the temperature of the surrounding air 36° F. resulted in an effect less than that which often took place when the natural change was only 4° F., the conclusion is that temperature changes cannot be regarded as the direct cause of these wanderings, although some effects may be due to alterations in temperature in the vicinity of the supporting column.

An experiment which promised to throw light upon the cause of these movements was to quickly empty a well 104 feet distant from a pendulum station of about two tons of water. This produced a tilt of 1″·36.

The direction of motion corresponded to that which would follow the removal of a load upon the well side of the instrument, a fact which suggested the establishment of a tide gauge in an unused well, eighty yards distant from the underground chamber, and sixty yards distant from the nearest well from which water was drawn.

The diagrams showed *two* sinkings and two risings, each about 5 mm., in the twenty-four hours. The sinkings took place between 2 and 6 P.M. and 2 and 5 A.M. Neither the double wave nor the general rising or falling of the water in the well showed any connection with the displacement of the pendulum.

Another suggestion as to the cause producing slow displacements which are not due to actual rock movement is to suppose that they result from the removal of a load from one side of a pendulum greater than that which is removed from the opposite side.

The effect of a load composed of men and boys standing on one side of a pendulum station is to depress the ground on which they stand, and the boom of the pendulum swings towards their side. When the load is removed, the ground rises and the pendulum returns to its normal position. A similar action may possibly be produced by the sun and wind carrying off more moisture from the ground on one side of a station than it does from the other side, the difference being due to the difference in the general character of such areas.

Inasmuch as the area which lost the greater load would have the greatest capacity for receiving moisture from time to time, a retrograde motion would be established, and during a year it would be expected that there should be an irregular wandering, first in one direction and then in the opposite direction, but it would also be expected that pendulums in different localities might behave differently.

This latter conclusion, although opposed to the inference to be drawn from the statement of M. Plantamour that the eastern piers of transit instruments in Europe rise during the summer, closely accords with the writer's observations in Japan.

The experiments which have been made in connection with evaporation and condensation will be referred to in a more detailed manner in a suggested explanation of the diurnal wave.

Seasonal changes in the loads carried by neighbouring areas will follow the appearance and disappearance of vegetable covering. In the case of thick grass or meadow land the seasonal difference in load may exceed seven tons per acre, and it is difficult to suppose that in the case of forests consisting of deciduous trees this quantity could be less. Another disturbing influence may occur at stations on sloping ground during a season of rain, the greatest load accumulating on the valley side of an instrument. An action of this description is very marked at Shide, in the Isle of Wight, where a pendulum oriented N. and S. invariably creeps at the time of rain towards the valley on the west. In fine weather the motion is eastwards.

Between October 1894 and January 1895, two pendulums, installed upon the surface of the alluvium in Tokyo at a distance of 416 yards from each other, showed

from curves of their midday positions that their wanderings, though unequal in amount, had followed the same
general directions, with periods of from four to about
thirty days. Local earthquakes were frequent during
twelve days of strongly pronounced westerly motion.

By comparing these curves with a similar curve showing changes in barometric pressure, it is seen that although
they agree in their general character, they do not closely
agree in detail. For example, six decided fluctuations,
each of from three to eight days period in pressure, correspond to *seven* sharply marked sinuosities on the curve for
tilting, which means that the relationship in direction of
motion of the barometer and the pendulum is at times
reversed.

The only pendulum in Japan which has been continuously observed without any change being made in its
installation is one which stands on a concrete floor in the
cellar of the Imperial College of Engineering. Its movements, like the levels of M. Plantamour, show an annual
periodicity. During the warmer months of the year it
creeps towards the west, whilst in winter this movement is
reversed.

The actual movements were as follows (see fig. 48) :

Sept. 1894 to June 11, 1895 . .	16·0″ East side	sinking.
June 11, 1895, to Aug. 23, 1895 .	7·0 ,,	rising.
Aug. 23, 1895, to Nov. 24, 1895. .	11·0 ,,	sinking.
Nov. 24, 1895, to Jan. 27, 1896 . .	6·0 ,,	rising.
Jan. 27, 1896, to Feb. 29, 1896 . .	2·5 ,,	sinking.
Total change	16·5 ,,	sinking.

Although the evaporation and condensation of moisture
may play an important part in slow periodic changes in
the vertical, many observations have been made which
lead to the conclusion that the soil on sloping ground has
a tendency to move slowly towards a lower level. Such
glacier-like movement might, under certain conditions,
result not only in the slow displacement of a foundation,
but in an alteration of azimuth and verticality. An indication of these movements is found in the shode stones

from the disintegrated portion of a lode which has its outcrop upon a hill side. From the position in which such stones are found, they would seem to have travelled through the alluvium from their origin in a downward direction, but along a line which brings them to the surface at some distance below their origin. More marked than these are the movements which have caused displacements in lines of railways running round and along the face of steep slopes. Sir William van Horne, writing

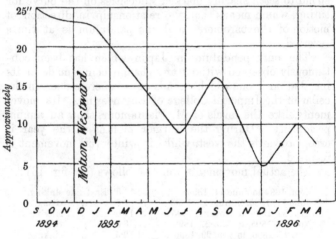

Fig. 48.

to me about the creeping of the earth which his engineers encountered in British Columbia, says that these movements have clearly been traced to the percolation of waters which had a natural source or came from ditches cut for irrigation. When, by sluicing, the water was diverted, the creeping stopped.

The movement is analogous to the downward motion of a sheet of lead upon a roof. In the case of soil resting on a slope, the penetration of moisture affects the usual conditions of cohesion and volume. By heat or frost there may be expansion, but whatever movement results from these actions under the influence of gravity, the displacements downwards are greater than the displacements upwards. The bearing of this upon the gradual but continous displacements observed by Plantamour, and upon the selection of a site for an observatory, is obvious.

Conclusions.—The preceding notes indicate that slow changes in the vertical are probably in operation in all localities. At different places they usually vary in their amount and in their rapidity, whilst their origin may be sought for in a variety of causes which act singly or together.

That movements are more pronounced parallel to the dip of certain strata rather than along the strike, and that the largest movements seem to be those which have been noted in countries where, for geological reasons, it may be assumed that mountain growth has not yet been completed, are facts which suggest that astronomers may already have measured movements which culminate in producing the striking features of the earth's surface. That earthquakes have been frequent when horizontal pendulums have been rapidly moving from what was apparently a normal position, raises the hope that clear observations of the change in the vertical may, in certain localities at least, furnish a warning of the approach of critical periods in rock bending.

The possible connection between the indications of a level and varying differences in load upon the two sides of a building in which it is placed, due to differences in the accumulation or evaporation of moisture, or seasonal differences in the vegetable covering on their sides, although effects due to such causes may be small, points to the advisability of obtaining for an observatory uniformity in environment. On a hill side change of level may be due to the accumulation of moisture acting as dead weight in the valley below, or to a slow downward creeping of the soil, which may be partially overcome by good drainage.

Barometrical effects are apparently only marked upon
soft ground, and, like the effects due to fluctuations in
temperature, they are usually short and irregular in their
periods. Under certain conditions, seasonal changes in
these phenomena may operate and cause seasonal fluctuations in level.

REFERENCES

1. Local Variations and Vibrations of the Earth's Surface, by H. C.
Russell, F.R.S., B.A., F.R.A.S., R.S. of New South Wales, July 1885.
2. Note on M. Plantamour's Observations by Means of Levels on the
Periodic Movements of the Ground at Sécheron, near Geneva, by Charles
Davison, M.A., 'Phil. Mag.,' Feb. 1889. This contains a list of M.
Plantamour's original papers.
3. Recherches sur la Verticale, par M. Antoine d'Abbadie, 'Annales de
la Société de Bruxelles,' 5ème anni, 1881.
4. Report Relating to the Measurement of the Lunar Disturbance of
Gravity, written in the name of G. H. Darwin, 'British Association
Reports,' 1881-2. This contains references to the work of Zöllner,
d'Abbadie, Plantamour, Ellis, Nyren, Bouquet de la Grye.
5. Earth Tremors, Report by a Committee, ' British Association
Reports,' 1893. This contains accounts of work by Wolf, d'Abbadie,
P. T. Bertelli, Milne. The Bifilar Pendulum of Mr. H. Darwin, the
work and papers of E. von Rebeur-Paschwitz (by von Rebeur-Paschwitz).
6. Earth tremors, Report of a Committee, 'British Association Reports,' 1894. This contains further notes on Mr. H. Darwin's Pendulums, the earthquake of 1894 ; and ' Observations at Nicolaiew,' by Prof.
S. Kortazzi.
7. 'Reports to the British Association, 1893-7,' drawn up by J.
Milne.
8. Das Horizontalpendel von Dr. E. von Rebeur-Paschwitz, Nova
Acta des Ksl. Leop.-Carol. Deutschen Akademie der Naturforscher.
Band lx. No. 1. This memoir contains an epitomised account of sixty-
four books and papers bearing upon changes in the vertical.
9. Horizontalpendel beobachtungen im meridian zu Strassburg. Dr.
R. Ehlert, ' Beiträge zur Geophysik.' iii. Band, 1 Heft.

CHAPTER XIV

THE DIURNAL AND SEMI-DIURNAL WAVES

A diurnal change in level observed by Plantamour, G. and H. Darwin, and
by Russell in Lake George—The records of von Rebeur-Paschwitz in
Teneriffe, Wilhelmshaven, Potsdam—Observations at Strassburg and
Nicolaiew—The records from nineteen installations in Japan, on
rock, in alluvium, underground, and on the surface—Possible relationship between the daily wave and the evaporation and condensation of moisture—Miller's experiments on evaporation—The loading
of areas by dew and subsurface condensation—Stones as condensers
and radiators—The transpiration of plants—The observations made
in Japan and the Isle of Wight in relation to the suggested explanations—Influence of the moon—Effect of tides.

VERY many observers who have had occasion to record the
daily readings of an instrument susceptible to slight
changes of level have noticed that such changes have had
a daily periodicity.

In 1878 M. Plantamour noted changes of this description, which on April 20 of that year reached a maximum
of 17·75 seconds. The eastern end of the level he used was
highest about 5.30 P.M., and there was a gradual rising in
its mean diurnal position.

In 1879 the eastern end of the levels was highest
between 6 and 7.45 P.M., and lowest at a similar hour in
the morning. In a north and south direction the movements were rare, irregular, and feeble, the maximum
excursion towards the north being reached about noon.

Messrs. George and Horace Darwin observed that the
maximum elevation of the south took place about noon,
which is the converse of the observation made by
M. Plantamour.

In 1885 Mr. H. C. Russell, F.R.S., described the diurnal
changes in the level of Lake George, a body of water in

New South Wales twenty miles long and five or six miles broad. These changes, which were clearly marked on tide-gauge-like records, seldom exceeded half an inch, but a tenth of an inch could be detected, corresponding to a change in the vertical of 0·016 second. The direction of motion corresponded to a rise of the southern end of the lake during the day, and a fall at night.

Diurnal changes of two or more seconds of arc, as observed by M. Plantamour or M. d'Abbadie, were never recorded. The changes which were noted did not appear to be connected with the instrument, the wind, or the state of the barometer.

For a most carefully conducted series of observations, the results obtained from which were subjected to a close analysis, our thanks are due to the late Dr. E. von Rebeur-Paschwitz.

The primary object of von Rebeur's investigations, which were made with horizontal pendulums, was the measurement of the gravitational effect of the moon, but the most prominent feature in the resulting photograms was a well pronounced diurnal period.

The amplitudes were subject to considerable variation, whilst there were also variations in the hours at which the pendulum reached the extreme limit of its eastern or western excursion.

In Teneriffe, for example, the deflections occurred about two hours earlier in winter than towards the end of April, whilst at Wilhelmshaven the eastern elongation is earlier in spring and autumn than it is in midsummer.

At Potsdam there is no marked change.

The means of these observations are given in the following table :

Movements of pendulums	Teneriffe	Wilhelmshaven	Potsdam
Completion of easterly movement (E. sank)	4.0 P.M.	3.0 P.M.	5.0 P.M.
Completion of westerly movement (E. risen)	7.30 A.M.	5.0 A.M.	8.30 A.M.
Range of motion . .	0·40 sec.	2·2 sec.	0·49 sec.

A comparison between the range of motion, the maximum oscillation of temperature, the number of hours of sunshine and the estimated amount of clouds showed that the average daily motion observed at Potsdam and Teneriffe agrees very closely with the meteorological elements named. Still, there are exceptions, where large displacements have been noted upon cloudy days and small displacements on clear days.

At Wilhelmshaven, although a similar relation is indicated, it is not so well marked.

At this place, also, Professor Boergen took readings twice a day of the level of the meridian circle, the pier carrying which rises from a mass of sand, which forms a bed to a depth of 250 m. beneath the marshy ground round Wilhelmshaven. The result of a month's observation showed no appreciable difference.

At Strassburg, however, the readings of a water level attached to a pillar showed a general agreement with the indications of a pendulum.

Von Rebeur, whom I have closely quoted, concludes that the oscillation is to a great extent due to the thermal effect of the sun, but as to how this effect takes place is yet an open question.

Curves of a similar character have been obtained at Karlsruhe, Strassburg, Charkow, and by Professor Kortazzi at Nicolaiew.

At Potsdam the instrument giving these records was established in a cellar, below the east tower of the Astrophysical Observatory. The installation at Wilhelmshaven was in a cellar of the Imperial Naval Observatory, whilst in Teneriffe the pendulum stood on the cement floor of a laboratory on the eastern flank of an old lava stream.

At Strassburg an instrument placed in an east-west plane recorded movements that were much smaller than those recorded when the pendulum was placed in the meridian.

In summer the southern elongation was reached about 6 P.M., and the northern about 6 A.M.; at Nicolaiew these movements take place about four hours later.

A fact of some significance is the observation of
Professor Kortazzi at Nicolaiew, who found that the
diagrams from a pendulum were very like those from a
hygrograph placed in the same cellar. The conclusion
was that the column supporting the pendulum behaved
like a sponge drawing moisture from the air, and thereby
causing a change in the inclination of the instrument.
When the openings to the cellar were closed and the
pillar covered with a waterproof material, this effect disappeared.

Observations made in Japan

The discovery of the diurnal wave in Japan was, as in
other countries, an accidental occurrence.

It was first observed in the photograms from an
extremely light form of horizontal pendulum set up for
the study of tremors. References to these various observations are given in British Association Reports for
1892-7.

The instruments giving the most satisfactory records
of these movements were the long boom horizontal
pendulums, from which nineteen sets of photograms were
obtained.

The chief object of these numerous installations was
to study the diurnal wave in relation to varying environ-
ments.

The greatest sensibility given to an instrument was
such that a deflection of 1 mm. was produced by a tilt of
0·1 second, from which it may be concluded that with
instruments of greater sensibility a diurnal wave would in
all probability have been measurable at stations at which
in the present section such a movement is described as
non-existent.

Five installations at which the diurnal wave was
imperceptible were underground in caves excavated in
fairly solid rock. As might be anticipated, the daily
changes in temperature in such situations were extremely
small.

With two pendulums placed upon a concrete floor in an underground chamber, about thirteen feet deep in the alluvium, where changes of temperature were insignificant, the daily wave was pronounced, its amplitude sometimes exceeding that observed at a station about 1,000 feet distant, where the supporting column rose above the surface of the ground inside a room in which the daily changes in temperature were considerable. At all the other stations, where the instruments were carried on short brick columns rising from the alluvium inside wooden huts, daily waves were nearly always visible.

It was clear from these observations that the recorded movements were not directly due to effects accompanying change of temperature in the immediate vicinity of an instrument.

At some stations the diurnal waves were always very small, but at one station they were sometimes abnormally large, indicating a change in the vertical of as much as forty seconds. It was seldom that two pendulums completed their excursions, say toward the east, at exactly the same hour, and cases occurred of two pendulums moving in nearly opposite directions at the same time.

One marked illustration of this was the case of two pendulums situated on the opposite sides of a swampy valley. If we imagine the trees on the bluffs bordering the two sides of this valley to follow the motions of the instruments, these bluffs or their coverings may be described as performing a daily bow towards each other.

It was observed that the diurnal wave was usually large on days that were warm and on which there was much sunshine, and also that on those days when it was cloudy or when rain fell the waves were reduced in extent or entirely disappeared.

These latter observations suggested the idea that the diurnal wave might possibly be due to evaporation, which during the day removed, to be dissipated in the atmosphere, a greater load from one side of a pendulum station than from the opposite side. An effect of this description would be most pronounced when two areas to the right and left of the plane of a horizontal pendulum differed greatly in the rates at which they gave up moisture.

The movement of a pendulum consequent on such an action we should expect to be most rapid during the middle portion of a day, as, for example, between 8 or 9 A.M. and 3 or 4 P.M. In the case of a pendulum installed in the meridian on an area uniform in character, the slightly greater evaporation which would take place on its eastern side in the morning should cause the pendulum to move westwards. Some time after midday this movement would gradually cease, and towards evening a retrograde motion be established. The chief cause of this retrograde movement at night is, however, likely to be occasioned by the acquisition of loads in the form of surface and subsurface condensation, the ground which during the day had become the driest being the most absorptive during the night.

Experiment showed that a load of about 1,000 lb., composed of men and boys, at a distance of fifteen feet from a pendulum was sufficient to cause a deviation of 2 mm. of the boom of a pendulum in their direction, whilst the emptying of a well at a distance of 100 feet of about two tons of water caused the ground on that side to rise sufficiently far to cause a deflection of 6 mm. away from the side from which the load was taken.

Professor H. H. Turner, of Oxford, in conjunction with the author, made experimental determinations of the deflections produced by a load consisting of seventy-six men, standing in close and open order, at different distances from his observatory. At a mean distance of sixteen feet for close order the change in level was 0·34 second, whilst for open order at a mean distance of twenty-two feet it was 0·16 second. A load of 240 lb. within seven feet of a horizontal pendulum caused a tilting of 0·49 second, whilst 350 lb. placed on the east and then on the west side of the base of a massive pier gave differences in readings on its top of 0·16 second (see B. A. Report, 1896).

Experiments on evaporation showed that earth lost on fine days 4 to 5 lb. of moisture per square yard, but as Mr. S. H. Miller, of Lowestoft, has made more accurate and careful experiments than my own, I give the following figures as daily averages computed from monthly observations of evaporation in pounds per square yard from various natural surfaces :

		lb.
Soil, humus (July)	. . .	4·24
Water (July)	. . .	8·61
Forest (Spruce)	. . .	12·52
Grass red clover (May)	. .	15·61

These numbers are practically in the ratios 1 : 2 : 3 : 4. Also we see that, so far as this possible cause is concerned, the greatest deflection of a horizontal pendulum would take place at a station on one side of which there was soil and on the other side grass, the differential relief of load being about 12 lb. per square yard. This is equivalent to the removal of about one ton on the grass side from areas each of which measure thirteen by thirteen yards, a load quite sufficient to produce deflections often noted during daytime.

The retrograde motion of a pendulum which takes place during the night may possibly be due to the side of a station which during a day has lost the most moisture receiving by absorption a load greater than that received by the other side.

In considering the causes which may result in this unequal loading, we must remember that the movement which takes place during the night is usually less than that which has taken place during the day, there being, in fact, a creeping of the zero point.

The night load for which we seek is therefore somewhat less than that due to evaporation, and it may be produced by the following causes, the effects due to which would be pronounced when they act in conjunction :

1. The unequal precipitation of moisture from the atmosphere, or its condensation as it emerges from the

ground on equally exposed but differently covered areas on the two sides of a station.

In Japan I noticed that as a maximum a grass surface growing in a box might in this manner gain from the atmosphere about ¾ lb. per square yard.

2. The unequal subsurface condensation of moisture on two sides of a station.

It is a matter of common observation that when a stone or a board that has been lying all night upon the grass is turned over, its under side is wet. This phenomenon has engaged the attention of Mr. Aitken, of Darroch, whose investigations on the formation of clouds and dew are well known ('Proc. and Trans. Royal Soc. Edin.,' 1880 to 1895); but the whole question of subsurface condensation seems to deserve a closer study still.

During a hot day soil is perceptibly heated to a depth of about one foot. After sunset the surface of this is quickly chilled and in winter frozen. As aqueous vapour rises upwards towards the cooled layer it is condensed, and therefore on certain nights surface soil may gain in weight.

As the result of a specially arranged experiment to measure subsurface condensation the author found that superficial soil would sometimes increase in weight about 10 oz. per square yard, which is about one-eighth of that which during the day had been removed by evaporation.

It can readily be imagined that this action will take place in different degrees upon differently covered surfaces.

With a stony soil, such, for example, as is found upon the chalk downs of the Isle of Wight and Hampshire, the stones which take up and lose heat quickly at night time act as condensers for the moisture rising in the earth beneath them. Therefore to entirely clear a land of stones may so far impoverish it by desiccation that the value of the crops would be impaired.

Admitting the reality of subsurface condensation, then,

we should expect two contiguous areas with surfaces
differing in character and with a common subterranean
supply of moisture to gain unequally in weight, and it

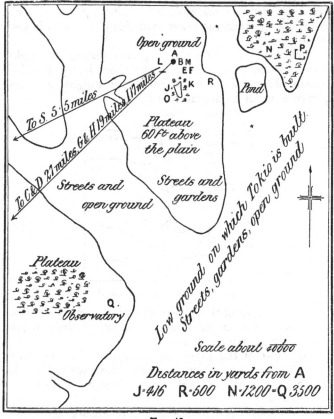

Fig. 49

seems likely that the area which during the day had
suffered the greatest loss would be the one to take up the
greatest quantity of moisture.

SEISMOLOGY

3. Different degrees of accumulation of moisture in neighbouring surfaces of different character. For example, if one surface was covered with a growth of plants and the other bare, then it seems likely that the former of these would, by virtue of the roots it contained, pump upwards more moisture than the latter, and this they would largely retain at night.

This is equivalent to stating that during the night, when transpiration is less active than during the day, certain plants may perhaps increase in weight in consequence of moisture drawn from below.

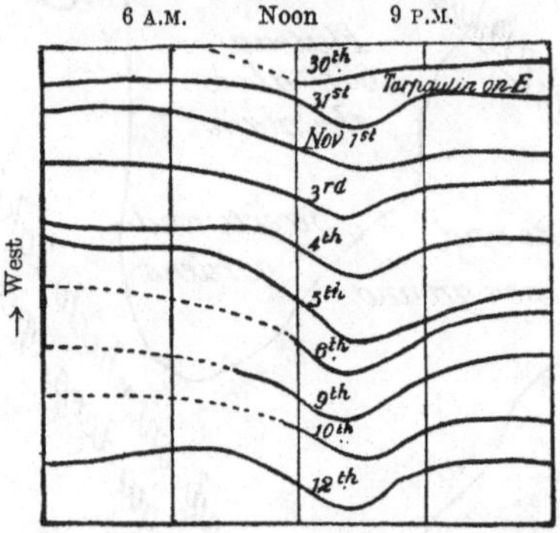

FIG. 50.—DIURNAL WAVES AT SHIDE, ISLE OF WIGHT, 1896

By one or all of these three means it seems reasonable to suppose that an area may at night partially regain weight lost during the day.

As a test of the suggested theory, we will now compare the results to which it would lead with the results obtained by actual observation from fourteen installations in Japan.

The stations are indicated alphabetically and their relative positions are shown in fig. 49.

Examples of the diagrams which have been obtained are shown in fig. 50.

The five installations in caves C, D, G, H, and I, upon rock did not, as we have said, show a daily wave.

Since above these excavations there was fifty or 100 feet of rock and earth, and since the instruments did not have a high degree of sensibility, the results observed are what might have been expected.

The instruments on the alluvium behaved as follows: four of these (A, E, J, and K), practically within a circle about 400 yards in diameter, were situated in the middle of a plateau about three-quarters of a mile wide and running from N.W. towards the S.W. to overlook the plain of Tokyo.

The daylight movement of two of these (J, K) was always westwards, or away from the most open ground and towards that most covered with buildings; the other two (A and E) usually moved westwards.

The exceptional movements to the eastwards may possibly be accounted for by the fact that their booms were not in the meridian, but oriented to point N.E.

Two other instruments (P and Q), on the eastern side. of two other plateaux had a daylight movement also towards the west, away from the open Tokyo plain.

With an instrument (N) upon the *western* side of one of these latter plateaux, overlooking an open marshy valley towards the west, the daylight motion was eastwards, or away from the area which might be expected to lose most by evaporation.

In considering what occurred at N, where the observations were made in November, December, and January, we must remember that the trees (cryptomeria) on its western side would, during winter months, lose extremely little by transpiration. On the contrary, they may possibly have absorbed moisture from the atmosphere.

In summer time they might give off more moisture than the open ground upon the west, and therefore the direction of motion which takes place during daytime would, at such a season, be reversed.

Lastly, on the flat plateau about five miles distant from Tokyo, at a station (S), the nearest irregularity to which was a small valley about half a mile distant, the daylight movement was also eastward, which was away

from a field of green corn and towards an open ploughed
field on the unfenced boundaries between which the
instrument stood.

In all these instances, therefore, the pendulums during
the daytime moved away from the side which might be
expected to be losing the greater amount of moisture.

One striking exception to this rule was the case of a
pendulum at (R), which was in the angle of a plateau so
situated that there was a deep cutting on its western side
and a steep scarp on its northern side. Very tall trees
overshadowed the hut, and the surrounding coverings on
the ground were very irregular. This pendulum, from
early morning until about 1 P.M., always moved in a rapid
but intermittent manner towards the east; from this hour
until 9 P.M. there was a western motion (fig. 56).

For half the day, therefore, the direction of displacement was contrary to that expected.

We therefore find that at thirteen out of fourteen
installations the movements usually, and in several cases
always, agreed with what the view just explained would
lead us to expect.

At Shide, in the Isle of Wight, the boom of an instrument installed on a tennis ground, which points from
north towards the south, and covered by a narrow hut
oriented in the same direction, moves but with a lag of
about two hours to keep itself in the same line as the
sun and the shadow of the hut. In the morning, therefore,
and up to about 2 P.M. the motion is rapidly eastwards,
after which the westerly excursion commences, and continues up to about 10 P.M. During the day, therefore,
the pendulum heels over to that side of the hut which
is the warmest, or in a direction contrary to that in
which we should expect it to move were its movements
dependent upon the removal of moisture immediately
round the hut. No change in the character of these
displacements was brought about by covering the ground
first on the eastern side and subsequently on the western
side with a large tarpaulin, the object of which was to

check local evaporation. It may also be noted that a trench six feet in depth and parallel with the western side of one of the installations in Tokyo failed to produce any marked alteration in the character of the diurnal wave. These observations indicate that we are not to seek for an explanation of the daily movements in differential evaporation effects, or the effects of unequal heating of the ground, in the *immediate* vicinity of an installation, whilst the results of observations made underground, where temperature variations are small, as contrasted with the results obtained at stations on the surface, where such variations are large, indicate that the causes of the movements are not to be found in the warping of the pier or portions of the instrument. The cause which results in a series of instruments on one side of a valley moving in the same direction at about the same time is evidently one that affects a considerable area. During a period of wet weather the booms of such a series of instruments slowly wander towards the side of the valley which is being loaded, whilst in fair weather the movement is in a contrary direction.

That there is a diurnal loading and unloading of a valley bottom apparently finds evidence in the diurnal fluctuation in the flow of certain rivers. During the day, whilst vegetable transpiration and evaporation are active, the volume of water received and carried by a stream is rapidly decreasing. At night time, when these activities are waning or have ceased, the stream slowly returns towards its former volume. Mr. Charles Hawksley showed the author a curve representing such changes, the character of which is closely identical with that of the diurnal wave.

Although so many of the observations which have been quoted strongly support the view that diurnal waves may be due to the suggested meteorological causes, there are many objections to such a conclusion.

For example, why should there be such marked differences in the amplitude of waves at different stations?

At J, K, and R the movements are large, whilst at

other stations, especially at N, they are comparatively small.

We must remember, however, that the side from which the greatest evaporation is believed to take place is determined somewhat vaguely from the general appearance of two sides of a station, and, moreover, that the side which loses the greatest load at one season may be that which loses the least at another season. It is, therefore, quite possible that errors may have occurred in making the necessary selections.

Another objection to the evaporation-condensation theory is found in the records from Potsdam, where to the east and west the ground is uniformly covered with pine, and yet the diurnal wave is marked.

Again, it is difficult to imagine how differences in evaporation and condensation at the extremities of Lake George could result in a daily tilting of the containing rocky basin. It must, however, be remembered that the diurnal changes at that place were extremely small, and may possibly belong to another order of movements than those observed upon land.

Another point is the smallness of the movements observed in a pendulum placed with its plane in the prime vertical, as compared with what is obtained from a pendulum placed in the meridian.

Before laying stress on this observation we must not forget that the installations of pendulums with booms in an east-west direction have been few in number, and that the movements of the east-west pendulum in Japan were moderate in their amplitude.

Admitting the gravity of these and other objections, and the fact that we have not established any relationship between the occurrence of the diurnal waves and such features as fluctuation in underground water, or expansions and contractions of the soil due to changes in temperature or dryness, we may nevertheless conclude that it is at least possible for this phenomenon to find a partial explanation in the causes which have been indicated. Should they at some future time be shown not to be the cause of the diurnal wave, they yet remain as causes that may lead to differential loading and unloading of neighbouring areas, and as such are worthy of the attention of those who seek sites for laboratories and observatories.

Dr. Reinhold Ehlert, who subjects a series of observations made between April and December 1895 to a careful analysis, concludes that the daily wave is due to a deformation of the earth's surface by the heat of the sun. The fact that in different months a pendulum in the meridian completes its excursion to the east or west at different hours strengthens the hypothesis. See ' Horizontalpendel-beòbachtungen in Meridian zu Strassburg,' i. E. von Dr. R. Ehlert, 'Beiträge zur Geophysik.,' iii. Band, 1 Heft. Also references to last chapter, Nos. 1, 2, 4, 5, 6 ; and ' Reports to the British Association,' 1892–7, by the author.

Influence of the Moon

The effects that the moon may produce upon the solid earth have been computed by Professor G. H. Darwin, who on this matter writes to me as follows :

' The various effects which the moon may exercise on a pendulum are very complex. First, as regards simplicity, is the effect of the force to which the oceanic tides are due. If the earth were absolutely stiff and unyielding, this tide-generating force would produce a periodic oscillation of the pendulum of an amplitude which can be calculated with a close degree of approximation. That amplitude is so small that the measurement of it, even by the most delicate instruments, is a matter of the greatest difficulty. But in the second place the moon's tide-generating force acts not only on the pendulum, but also on the earth ; and as the earth cannot be, as a whole, absolutely stiff, it must yield to the force. If it yielded as freely as water the earth's surface would necessarily be perpendicular to the pendulum, and the pendulum would remain apparently at rest. But it does not yield with perfect freedom, and therefore, in as far as it yields, its movement imparts to the pendulum an apparent deflection which tends to mask the true deflection due to tide-generating force. Lastly, at places within a few hundred miles of the sea, the varying load of the oceanic tide must produce a deflection of a pendulum, which is partly real and partly apparent. The real portion is almost certainly by far the smaller ; it is due to the direct attraction of the sea, which will vary in intensity with the alternations of high and low water. The apparent portion is due to the warping of the superficial strata by the varying load of the tide, the slope being towards the sea at high water, and away from it at low water. I suspect that where a lunar periodicity of the pendulum has been observed, it has been principally due to this warping of the superficial strata.'

Von Rebeur-Paschwitz, who alone may possibly have experimentally measured the lunar influence, says that if the difference which is observed in these quantities is attributed to a general elastic deformation of the earth, the maximum rise and fall of the surface will be ± 11 centimetres, and that the summit of the elastic wave is not below the moon, but precedes it by about two hours. The size of the wave as recorded at a given station will vary with the moon's declination, and show certain regular changes. The analysis of the records from Potsdam and Puerto Orotava gave some evidence of a small lunar wave of 0·01 second. The amplitude at Strassburg is 0·018 second. On the photograms from Wilhelmshafen, as on many from Japan, a semi-diurnal wave was distinctly visible. At Wilhelmshafen this takes place about half an hour before the meridian passage of the moon and one and a quarter hours before high water. It is, of course, possible that the displacements might be due to tidal depression, but if this were so then, as von Rebeur-Paschwitz remarks, there would be a close connection between the lunar terms and the form of the tides, which is not the case (see 'British Association Report,' 1893, p. 312 &c.).

The semi-diurnal wave observed in Japan may have been connected with the semi-diurnal rise and fall of water noted in an unused well (see p. 243.) Dr. R. Ehlert, like von Rebeur-Paschwitz, and also from observations made at Strasburg, finds evidence of a lunar effect (see references p. 263.)

Professor S. Kortazzi, who analysed the records obtained from a horizontal pendulum placed in the prime vertical at Nicolaiew, concludes that the influence of the moon is insensible. Assuming that an elastic tide is as marked as the analysis of Paschwitz indicates, then we should expect a more marked coincidence than has hitherto been shown to exist between the times at which earthquakes are frequent and certain phases of the moon (see ' The Tides ' by G. H. Darwin, F.R.S., 1898).

CHAPTER XV

PULSATIONS

Distinction between tremors or microseisms and pulsations—Identity
in the character of observations made in Japan, Germany, and the
Isle of Wight—The period and amplitude of pulsations—Pulsations
chiefly occur in winter—Relationship of pulsations to atmospheric
pressure and unusual oceanic disturbances—The cause of tidelike
ocean waves.

THE observation that, during a tremor storm with maxima
recurring at intervals of from four to eight minutes, two
light horizontal pendulums, placed side by side to record
the same component of motion, would commence to move
from positions of comparative rest at about the same time
and in the same direction, led me to the conclusion that
the column on which the instruments stood was subjected
to intermittent tilting, and that the so-called earth tremors,
rather than being due to elastic movements, were the
result of pulsatory disturbance of the earth's surface comparable to an ocean swell. Such movements succeeded
each other at intervals of two or three seconds, and the
maximum deflections varied between one and four seconds
of arc. Even if these movements had a seismic origin,
waves which might be five or six kilometres in length,
rather than being called microseismic, might be more
accurately described as megascopic. These movements I
have called earth pulsations. They seem to be more
closely connected with fluctuations in barometric pressure
over considerable areas of the earth's crust rather than
with causes endogenous to the earth's crust.

From time to time on the photographic traces groups

of exceedingly small but extremely regular waves showed themselves, which had periods of from two to three minutes. Many years ago examples of these were sent to the late E. von Rebeur-Paschwitz for comparison with his records made in Germany, and the conclusion arrived at was that we in Japan were evidently recording similar phenomena. The distinction between irregular intermittent tremors and these regular motions is undoubtedly a sharp one, and although there lie between the two groups others of varying degrees of regularity, I shall, for purposes of convenience, follow the suggestion of von Rebeur-Paschwitz, and confine

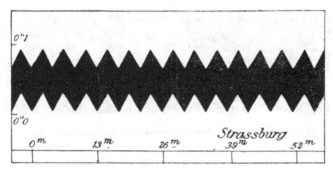

FIG. 51.—PULSATIONS AT STRASSBURG (PASCHWITZ) × 10

the term 'tremor' to the former and 'pulsation'[1] to the latter type of movement.

Fig. 51 is a reproduction, magnified ten times, of a record given by von Rebeur-Paschwitz from the observations carried out by Professor Becker at Strassburg. It may be regarded as identical in appearance with records from an extremely small horizontal pendulum used in Japan.

[1] In the author's volume on 'Earthquakes,' written in 1883, the expression 'earth pulsation' referred to large wavelike undulations on the surface of the earth which we now know are movements due to earthquakes which have originated at great distances from the place where the undulations have been observed.

The period of the waves shown is 3·25 minutes. Usually
the period varies between two and three minutes, but
waves with a period of one minute have been noticed.
When waves are superimposed on the records of large
movements, or where they have succeeded each other so
rapidly that they have formed knots along the otherwise
straight trace of a photogram, all that can be said is that
there is evidence of waves with still shorter periods, and of
comparatively short duration.   With these von Rebeur-
Paschwitz includes regular waves and periodical impulses,

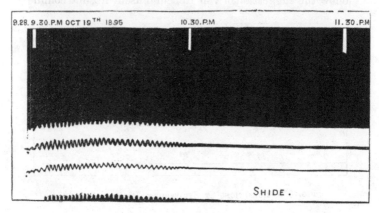

9.28. 9.30.P.M OCT 19 TH 18.95   10.30.P.M   11.30.P.M

SHIDE.

FIG. 52

which usually have a period of five or six minutes, but
which may reach fifteen minutes. I have also recorded
the same classes of movement, both in Japan and the Isle
of Wight, but they are of unusual occurrence. Although
waves of the character shown in fig. 51 may not have
an amplitude exceeding 0·05 second; fig. 52, which is
reproduced from a record obtained in the Isle of Wight,
shows waves with amplitudes of 1·42 seconds. The number
of waves in a disturbance where the period is two or three
minutes is usually from thirty to a hundred; when the

periods are from five to ten or more minutes, the numbers are reduced to between three and twelve.

The first records I obtained of waves having periods varying between a few minutes and one hour were obtained between January 1885 and May 1886 (' Trans. Seis. Soc.' vol. ix. p. 1–78).

Von Rebeur-Paschwitz recorded pulsations in Strassburg and at Teneriffe which, in their regularity and time of occurrence, were closely related to each other. The fact that they are almost confined to the autumn and winter (middle of October to middle of January) may possibly explain the reason why they were not observed at Potsdam and Wilhelmshafen. An extremely curious feature presented in the table of observations made at Strassburg is the fact that in October the pulsations commence with a period of a little over three minutes, which gradually and with fair regularity decreases to about two minutes in the middle of January. The result is that any group of six or ten successive disturbances have approximately the same period. These disturbances, which number sixty-three in all, may be divided into thirty-six represented by 1613 waves which occurred between 6 P.M. and 6 A.M., and twenty-seven with 1,450 waves which occurred between 6 A.M. and 6 P.M.

The Japan records for 1885 show that waves of from one to thirty minutes period were observed ten times in January and seven times in February. Between May and August slight waves only occurred four times. This would indicate a winter rule similar to that observed at Strassburg. The Strassburg records also show that pulsations have been frequent at about the time of new moon, but although this suggests that their appearance may be connected with oceanic tides, it seems impossible to imagine that such influences should be apparent only during the winter months. The occurrence of pulsations does not appear to be connected with any ordinary changes in temperature or atmospheric pressure, for if such a connection existed then the regular movements which pulsations exhibit should not be so markedly absent during the summer.

In the chapter on tremors it will be seen that these movements, which at times closely resemble pulsations, may possibly find an explanation in the intermittent application of barometrical loads over districts the parts of which have varying degrees of elasticity. If pulsations are regarded as exceptional forms assumed by tremors, then they may result from exceptional barometrical conditions.

Such an explanation is, however, extremely unlikely. Attributing the slight differences in period which are exhibited in a set of successive waves to the viscosity of the moving materials, we then have to account for the establishment and maintenance in a portion of the earth's crust of a free period which is very much slower than we might anticipate. On the contrary, if the differences in period, which in fig. 52 varies between six minutes twenty-seven seconds and two minutes seven seconds, are not negligible, the movements, although nearly isochronous, are of a forced character.

Phenomena which at first sight might appear to be connected with these earth movements, and at the same time more potent than fluctuations in barometric pressure, are sudden oceanic disturbances, as, for example, those which are noted upon the coasts of Peru and Japan.

At Callao and Paita the highest tides prevail in December and January, and they are accompanied by enormous waves or sea swells, which from time to time are thrown upon the coast. They vary in their duration from twenty-four to twenty-seven hours, and are accompanied by an unusual height of tide. They are not connected with atmospheric storms or with wind, and they occur, but not always, about full moon. They are annual and constant in their periodicity, and most noticeable between Tumbez, 3° S. lat., and the Chincha Islands, 14° S. long.[1]

---

[1] 'Notes on Tides at Tahiti,' *Amer. Jour. Sci.* vol. xlii. p. 45.

On the east coast of Japan what may be a similar phenomenon appears about the end of September.

The annual recurrence of these waves and their period preclude the idea that they have a seismic or volcanic origin.

Their effect upon a coast line would be to produce a bending, the statical effect of which would not be recognisable at any great distance; but the quick removal of these loads might possibly result in rhythmical deformations which would be propagated inland. Accepting, but only for the moment, this unlikely explanation, we should expect to find that near a coast line the amplitude of this pulsation would be larger than those recorded at a station many miles inland. On October 19, 1895 (fig. 52), this appears to have been the case, the corresponding record at Strassburg being an exceedingly small broadening of the record, as if produced by tremors.

This, however, being the only instance of a coincidence between the times at which pulsations have been observed at distant stations, is an observation carrying but little weight. Moreover, on several occasions when pulsations have been recorded in the Isle of Wight, at about the same time the charts from tide gauges on the neighbouring coast have not shown any marked irregularities in their traces. The idea that pulsations may be connected with oceanic disturbances is therefore one that receives but little support from facts, and to speculate as to their origin, beyond considering them a possible form of ' earth ' tremors, is until they have been more closely observed a task hardly likely to increase our knowledge.

The fact that a delicately adjusted horizontal pendulum is always in motion suggests the idea that some of the pulsations may find an explanation in the hypothesis that within the cases of these instruments there are currents of air produced in a manner similar to that referred to in the chapter on Earth Tremors.

CHAPTER XVI

EARTH TREMORS

General character of tremors—Distinction between tremors and pulsations—Observations of Bertelli and Rossi—Results obtained in Japan
—Relationship between tremors, wind, barometric gradient, the rate
at which atmospheric pressure changes, and waves upon a coast—
Observations of M. d'Abbadie—Tremors and earthquakes—Tremors
in relation to the hours of the day—Observations of von Rebeur-
Paschwitz—Tremors, wind velocity, frost, and the diurnal wave—
Artificial production of tremors—The character of the record changes
with the instrument employed—Air currents in closed cases produced by desiccation—Tremors probably due to changes in barometric
pressure, and expansions and contractions of the soil.

THE movements known as earth tremors or microseismic
storms are those which from time to time announce their
existence by unequal and fitful disturbances in pendulums
and delicately suspended apparatus of a like description.

When tremors do not exist, and a light but short
horizontal pendulum is caused to swing, it does so with a
regular period, but when it is moving under the influence
of tremors it moves with an irregular and variable period.

At times this irregular motion may cease, but a few
moments later the swinging will recommence, and periods
of maximum motion are reached every few minutes.

A tremor storm nearly always commences gradually,
and takes several hours to attain any marked intensity.

When watching an instrument moving under the
influence of tremors we gain the impression that the pier
on which it stands is from time to time subjected to
irregular tilting. When these movements agree with the
period of the pendulum, they result in maximum displace-

ments, whilst when there is a failure in synchronism there
is a tendency to produce rest.

When tremors are recorded by a long horizontal
pendulum adjusted to have a period of about fifteen
seconds, they show themselves as irregular and intermittently performed back and forth movements, which on
a photogram are seen as long period waves more or less
serrated in their outlines.

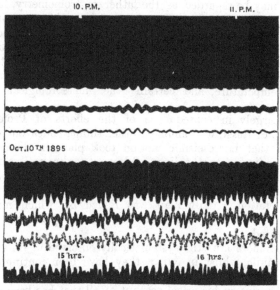

FIG. 53.—COMMENCEMENT OF A TREMOR STORM AT 10 P.M. AND
ITS CONTINUANCE AT 3 A.M.

When the period of the pendulum approaches one or
two minutes, if this period remains constant and the waves
are smooth in outline, the resulting record is that of the
so-called earth pulsations, to which certain tremor storms
have a close resemblance. Not only must they be distinguished from these, but they must not be confounded
with minute vibrations of an elastic character, like those

T

produced by a passing train, to the effects of which most
horizontal pendulums are practically insensible.

A tremor storm has but seldom a duration extend
ing only over three or four hours; usually it lasts from
eight to twelve hours, but it may continue for two or
three days.

In Italy, where microseismic motion has been more
extensively studied than in any other country, Bertelli,
who may be regarded as the father of tromometry, found
that tremors were more frequent in winter than in summer,
that tremor storms were closely associated with periods of
barometric depression, that periods of great activity
occurred at intervals of about ten days, and that these
disturbances have no connection with wind, rain, change
of temperature, and certain other meteorological conditions.

Largely in consequence of the efforts of Professor
M. S. di Rossi, who amongst other things endeavoured to
show that microseismic motion took place in directions
perpendicular to the lines of known faults, no less than
twenty-seven stations for the observation of tremors were,
in 1887, established in the Italian peninsula. At a central station a daily map was issued on which micro-
seismical activity throughout the Italian peninsula was
indicated.

Tremors are still systematically observed in Italy and
in Manila, whilst in Japan they have been continuously
recorded for many years with a variety of apparatus.

To give an historical account of all that has been done
towards the investigation of these ubiquitous movements,
especially in the Italian peninsula, would be entirely
beyond the scope of the present chapter. All that is
therefore attempted is to give an outline of the general
results which have been reached respecting the times and
conditions at and under which microseismic disturbances
have been recorded.

What follows is for the most part based upon my own
observations in Japan and the Isle of Wight, the records

having been obtained with apparatus giving a continuous photographic record.

In these countries, as in Italy, tremors appear to be most frequent during the winter and at such times when the barometer has been low, but there was no evidence of a ten days decadic periodicity.

One important observation was that tremors were nearly always at a maximum when the barometric gradient was steep, no matter whether at the place of observation the barometer was high or whether it was low.

This relationship between the occurrence of tremors and atmospheric conditions seems to destroy the distinction drawn in Italy between storms called *baroseismic* motions, which appear at the time of a low barometer, and the *volcano-seismic* disturbances, which occur at the times of high pressure.

Although it may often happen that the velocity of the wind is far from being proportional to an existing gradient, an inference to be drawn from the relationship just established is that tremors should be frequent when a strong wind is blowing, although wind might not be noticeable near the observing station.

The correctness of this view has been confirmed by comparisons which have been made between tables of tremor records and weather maps.

For example, in 1887 strong winds were blowing in Central Japan eighty-six times, and it was only in six of these cases that tremors were not observed. On the contrary, when it was calm in Central Japan, only extremely small tremors were noted, and even this was of rare occurrence.

An analysis of tromometric records from Rome showed that for sixty-three increases and decreases in microseismic intensity, there had been forty-six increases and decreases in the intensity of the wind ; the corresponding numbers for Rocca de Papa being sixty-four and forty-six.

Another investigation has revealed the fact that whenever there is a barometrical change of six or more than

six millimetres in eight hours (which usually occurs with
a falling barometer), tremors are pronounced.   Earth
tremors may therefore be connected with the rate of change
of pressure.

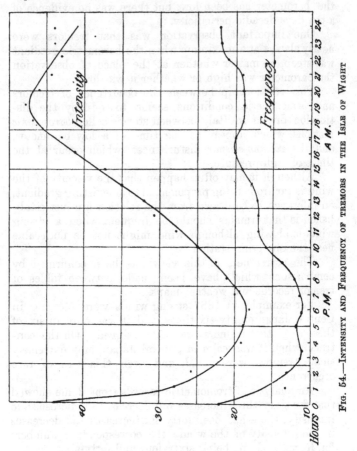

FIG. 54.—INTENSITY AND FREQUENCY OF TREMORS IN THE ISLE OF WIGHT

It was observations of this nature which led me to
consider wind, even when acting at a distance, to be, by

reason of its mechanical action on the irregular features of a country, one of the principal causes producing tremors.

Although wind may modify a tremor record, the observation, by no means uncommon, of a tromometer being perfectly at rest whilst a heavy gale was blowing round the observatory, shows that the connection between the two sets of phenomena is not so close as might at first be supposed.

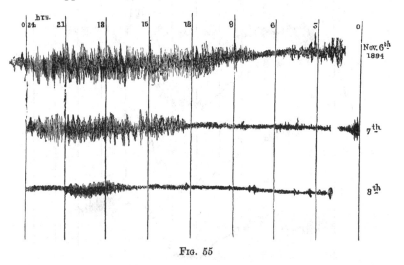

FIG. 55

Further than this it ought to have been an *a priori* conclusion that vibrations established by wind would be of an elastic character, and therefore hardly likely to produce movements like those observed.

M. d'Abbadie explained the 'movements' on the surface of the mercury in his 'naiderine' as being due to the beating of the waves upon the adjacent coast, but no such connection was observed between tromometer records in Tokyo and the violence of waves at distances of fifty-three and thirty-three miles.

The effects noted by M. d'Abbadie were in all probability movements similar in character to those which result from the action of wind, to record which ordinary tromometers are not adapted.

M. S. de Rossi and other Italian observers have

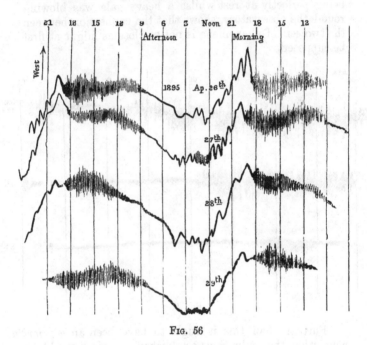

FIG. 56

pointed out some remarkable instances where tremors have been the precursors of earthquakes.

This does not appear to have been the case in Japan.

One curious feature in connection with the occurrence of tremors is their relationship with the hours of day and night.

Both in Japan, the Isle of Wight and in Toronto they have been more frequent during the night than during the day, and their frequency and intensity has usually reached a maximum about 6 or 9 A.M. (fig. 54.)

With a tremor storm continuing over several days, maximum motions also occurred about these hours.

As the maximum died away, movements would only be observed at night, and the last seen of the storm would be slight motion somewhere about the time of sunrise.

An extinction of this description is shown in the records for three successive days in fig. 55.

At other stations where tremors were seldom observed traces of these might appear about the time of maximum at stations where they were continuous for several days.

At one station tremors appeared almost every night, commencing about 9 P.M., ending at 7 A.M., with a maximum about 5 A.M. (see fig. 56).

Von Rebeur-Paschwitz found from careful examination of the tremor records obtained at Strassburg that there was a very close connection between the occurrence of tremors and the velocity of the wind.

The intensities of these two quantities were not, however, always proportional to each other, especially in January and December, when the tremors reach a maximum. This may be attributed to the fact that during winter the ground, being frozen, is better suited for the development of these movements.

He also noted a distinct daily period in the occurrence of tremors, which occur more frequently during the day than during the night.

These times of maximum are approximately as follows :

Spring (March—May)	4 P.M.
Summer (June—August)	2 P.M.
Autumn (Sept.—Nov.)	10 A.M.
Winter (Dec.—Feb.)	2 P.M.
For the year	3 P.M.

These results, it will be observed, are directly opposed to those obtained in Japan, the Isle of Wight, and Toronto.

In connection with the winter frequency, I may mention that a light horizontal pendulum in the Isle of

Wight shows marked tremors or movements of a very pronounced character every frosty night, but on nights when there is no sudden fall in temperature it remains at rest.

Returning to the observations made in Japan, it will be noticed that tremors are most frequent whilst a horizontal pendulum is moving most gradually, but that they are pronounced about the time when the fairly sudden reversal in the direction of its movement takes place.

The fact that tremors were produced at the time a well was being gradually emptied of water (p. 243) suggests the idea that either a pendulum takes up its position intermittently, or that the level of the ground is changed intermittently.

Whichever it may be, one conclusion is that whenever a rapid change in the inclination of the ground takes place, horizontal and possibly other forms of pendulums may be caused to swing.

It also seems that there may be a close relationship between times at which microseismic motion is pronounced and the escape of fire damp at certain collieries, a matter which, on account of its practical importance to miners, is specially referred to in the next chapter.

Considerable light seems to be thrown upon the cause of earth tremors when we compare their occurrence with the character of the tromometer which has been employed and the conditions under which it was observed.

In Japan, tremors have never been observed underground in caves, although they have been noted in an underground but fairly ventilated chamber in the alluvium. At Rocca de Papa in Italy, however, tromometers installed in caves have shown movements like those observed upon the surface.

The pendulums which yielded the most pronounced records of tremors were those which were actually the lightest, or the lightest relatively, to the length of their booms. For example, pendulums the booms of which varied from one quarter to three or four inches in length, were often in movement.

Many tremor records were obtained from a pendulum established at R, see p. 257 and fig. 56. The boom was about two feet six inches in length, and the instrument behaved in a very similar manner when brought to the Isle of Wight.

There is no reason to believe that any of these instruments were affected by air currents entering the covering cases from the outside, or that air currents were established inside them in consequence of differences in temperature of their different parts.

Tremors have been just as marked with pendulums covered by a case completely lined with very thick felt, as when the internal walls of the casing were partly of wood and masonry.

Inasmuch as the tremor records from the instrument in the Isle of Wight have been greatly reduced in numbers and intensity by surrounding the light boom with a shield against air currents, it may be inferred that such air currents are from time to time brought into existence. A difference in the rate at which moisture is absorbed, condensed, or evaporated in different portions of a covering might possibly account for such currents, and the fact that a light pendulum may be caused to swing by introducing a tray of calcium of chloride beneath its case supports this supposition.

The instruments on the alluvium which showed the least movement, and at some of which tremors were never observed, were those covered by a well ventilated light wooden hut and an imperfectly fitting case. It was difficult to understand why these instruments had never been set in motion by air currents from the outside.

Instruments on the best foundation and in every way well protected from outside influences often showed excessive movements.

One possible cause for these currents is that they may accompany the inflow or outflowing of air through fine passages, or even an osmotic action consequent upon difference in pressure or temperature between the outside or inside of the case. As a matter of fact the whole character of a tremor record may be changed, or even the tremor movements be brought to an end, by opening the covering of a case for a few moments and then closing it.

Well ventilated instruments, it has been remarked, seldom showed tremors.

Although the fact that tremors were most marked in the records from light pendulums beneath closely fitting cases supports the view that internal air currents may give rise to some of the movements called earth tremors, yet such an explanation cannot account for all.

For example, we have to explain why such currents are at certain stations only established between 6 A.M. and 9 A.M. and during the night, why they have a winter maximum of frequency, and why they are affected by the meteorological conditions already formulated. A still more difficult point for a theory of air currents to explain is the manner in which a pendulum repeatedly heels from side to side two or three times more slowly than if it were swinging naturally.

One explanation of earth tremors refers them to barometrical pressure and the rapidity of its changes.

In 1885 in Central Japan, where my own observations were chiefly made, the gradients expressed in millimetres per 120 geographical miles were as follows:

January	.	. . 6	July .	. . .	. 0
February	.	. . 5·6	August	. . .	. 0
March .	.	. . 5	September	. .	. 0
April .	.	. . 1 to 4	October	. . .	. 2
May	.	. . 0	November	. .	. 8
June	.	. . 0	December .	. .	. 8

This table shows a close accord between the times at which tremors were pronounced and gradients were steep. The steeper the barometric gradient across a district, the more likely it is that tremors should occur, and in Japan it has invariably happened that where there has been a gradient of 6 mm. per 120 geographical miles tremors have been recorded.

A more important observation, however, is that these movements have accompanied rapid changes in pressure, as, for example, a fall of 6 mm. within a period of eight hours.

The statical effect of a barometrical grade of 5 cm. in 1,500 miles over a tract of the earth's surface with an assumed rigidity of 3×10^8 (grms. per sq. cm.) has been estimated by Professor George Darwin, who finds that the ground under the depression would be 9 cm. or three and a half inches higher than under the elevation.

This gradient is four times greater than any gradient noted in Japan, but the rigidity modulus, which is that of glass, is also very much higher than that of much of the material met with upon the earth's surface. The deflection estimated by Darwin, which is approximately 0·006 second of arc, may therefore be taken as being less than that which we should expect to find under natural conditions.

Both in Japan and Germany slow changes in position of a horizontal pendulum have been recorded as accompanying changes in barometric pressure.

At Wilhelmshafen von Rebeur-Paschwitz found that a change of 1 mm. in the barometric pressure corresponded to a change of 0·29 second in the vertical.

Inasmuch as these measurable deformations of the earth's surface are produced by barometrical loads, it does not seem unreasonable to suppose that, when these loads are moving, not uniformally but intermittently, with average velocities exceeding twenty geographical miles per hour over a district which is not uniformly elastic, the result might be to break the surface up into irregular waves.

Different districts would be affected differently, whilst the effect in a given district would vary with the form of the isobars which crossed it, the direction and rapidity of their movements, the steepness of the gradient, and a number of other meteorological conditions.

A strong objection to this suggestion respecting the origin of earth tremors is that a surface having a high degree of rigidity would quickly adjust itself to the effects of load travelling at the supposed rates, and that barometrical records but seldom show the rapid alternations in pressure which are apparently required.

What has been recorded, and what is not uncommon, are changes in pressure of one or more millimetres per hour, which may represent the addition or removal of loads of from 1 to 2 lb. every two minutes on each square yard of surface.

If these changes take place intermittently and gusts of wind are indications of rapid changes in pressure, small as they are, it seems possible that they might be sufficient to establish motion in instruments susceptible to slight changes in level; but even if they are fairly uniform, the action taking place upon surfaces offering varying degrees of resistance to compression and with varying resilience, might result in differences in rate of yielding and in the movements called earth tremors. Should this be the case, tremors would have a greater frequency during frosty weather, there being at such times a greater difference in resistance to yielding, owing to inequalities in the hardening of neighbouring areas by freezing.

Tremors are at certain installations always pronounced on frosty nights and usually continue whilst the ground is thawing under the influence of the morning sun, a fact which suggests as an explanation for such a coincidence that at these times the ground around the observing station is subjected to comparatively rapid expansions and contractions. On nights when there is marked transpiration of moisture from soil and plants, as evidenced next morning by a copious dew, tremors are also frequent. The fact that they are frequent when there is a fall in temperature, although it suggests the idea that with these changes there has been an inflow or outflow of air from the covering cases of the instruments, does not similarly explain the movements observed in delicately suspended apparatus beneath cases which are airtight. For the explanation of this phenomenon we look to convection currents.

Conclusions.—As the result of observations similar to those outlined in the preceding pages, it would appear that the causes of the so-called ' earth ' tremors are two-fold.

1. Air-currents within the cases. Such currents are produced by a cold current of air impinging upon the outside of covers like glass or thin metal, but they are not likely to be produced if the covering is made of thick wood lined with thick felt. They may be produced by an inflow or outflow of air through ill-fitting joints, but what is more likely, as experiment has shown, by a difference in the rate at which moisture is condensed, absorbed, or given off at different points within a cover.

In many instances the circulation thus established within a case might be expected to be slow and uniform. This, acting upon a light pendulum with a natural periodic motion, would give rise to a resultant movement which over considerable intervals of time would be periodic in character. In this way it does not seem to be entirely beyond the limits of possibility that regularly recurring forced displacements having the appearance of earth tremors, and even pulsations, might occur.

2. By movements in the superficial soil outside the building in which the instrument is installed. These movements take place in soil whilst it is freezing or thawing, and after a heavy shower on dry ground. They may also be produced at a time when there are rapid but small changes in barometric pressure over an area the different portions of which vary in their elasticity and resilience.

Although these suggestions partially destroy the value of many records of ' earth ' tremors, they nevertheless leave us confronted with phenomena which it is the interest of all who have to work with instruments having delicate suspensions to understand more clearly, especially, perhaps, the reason that their frequency is so marked at particular hours and seasons.

CHAPTER XVII

MOVEMENTS OF THE EARTH'S CRUST IN RELATION TO PHYSICAL RESEARCH AND ENGINEERING

Bradyseismical motion in relation to harbour works—Cadastral surveys —Changes in the height of hills—The creeping of soil—Diurnal waves in relation to agriculture, forestry, and physical investigation— Behaviour of balances—Earth movements and astronomical observations—Fire damp and earth tremors—Fire damp and the barometer —Observations at collieries in Japan and France—Artificially produced vibrations—Effects produced at Greenwich—Prof. Paul's observations at Washington—Effects of sound waves on buildings— Vibration of railway trains, bridges, buildings, steamships.

THE radical changes and new principles that have been advantageously introduced into engineering and building practice in earthquake-shaken districts strengthen the hope that the extension of our knowledge relating to naturally and artificially produced vibrations and earth-movements in general may also lead to results not altogether devoid of profit. Bradyseismical movement is all-important to the geologist, but is usually considered to be of little interest, if any, to the practical engineer and surveyor. During a period of sixteen years the ground in the neighbourhood of the well-known Jupiter Serapis is said to have sunk at the rate of one inch in four years, whilst at Yokohama, if we had the means of accurate measurement, it is not unlikely that an elevation at several times this rate might be determined. In seventy-two years it is therefore within the limits of possibility that the harbour now in construction at Yokohama may be a fathom shallower than it is designed—

a rate of change, however, which with a muddy bottom is well within the control of dredges and other engineering appliances. With this rate of change, which is comparable with the rate estimated by Mr. Darwin for Valparaiso, the future of harbour works with rocky bottoms, the levels of canals and lines of water pipes, might be so far altered by bradyseismical action that, although not of great importance for the living, such changes might be yet worthy of consideration.

To what extent cadastral surveys are slowly losing accuracy, through the compression of valleys and the growth of mountains, we have no definite knowledge ; all that we do know is that when these movements are completed suddenly, or after they have gone on gradually for a long period of years—as was the case around Niigata on the western coast of Japan—re-surveys are required.

Comparatively rapid changes, resulting in differences measurable after intervals of a few days, in the height and even the position of hills, although by no means unknown, are so infrequent that their consideration is of small importance.

Changes of this character, when they have taken place with comparative rapidity, may in some instances be attributed to the sudden and by no means uncommon displacements called landslips.

The creeping of the soil down steep slopes has been very forcibly brought to the attention of engineers of the Canadian Pacific Railway, who in the gorges of the Rocky Mountains have seen the railway tracks slowly changing their original directions.

One explanation of the slow but continued change in inclination observed by Plantamour is that it may be due to the steady creeping of the soil upon a comparatively gentle slope, a suggestion which, if true, is not without importance to those whose mission it may be to select a site for an observatory.

In previous chapters reference has been made to the numerous new rules and formulæ introduced into the practice of engineers and builders whose object is to mitigate the effects of earthquake motion. That the new departures in engineering and building practice have proved beneficial has now been repeatedly demonstrated, and we feel assured that in Japan at least tall chimneys, factories, bridges, ordinary dwellings, and other structures embodying the new methods may be severely shaken and remain intact, whilst ordinary structures would be badly shattered or collapse.

In these matters seismological investigation has done much to reduce the loss of life and property in earthquake districts, and in recognition of what has already been accomplished and what yet remains to be done the Japanese have established a Chair of Seismology at their University, and appointed a committee to make investigations respecting earthquake effects, and a seismic survey of the empire.

In 1891 Japan lost 10,000 lives and incurred an expenditure of at least 30,000,000 dollars in repairs ; the same country in 1896 lost nearly 30,000 people, which is a number approximately six times the number killed in the recent Japanese-Chinese war ; and every year in some country or other where British and other capital is invested disastrous earthquakes take place ; all such calamities demonstrate the great importance of investigations whose aim is to mitigate their dire effects.

By the use of seismographs along the coast of Japan submerged areas of seismic activity have been mapped through which it would be dangerous to lay a cable. Instruments which record the unfelt movements of the earth's crust tell us about suboceanic geological changes, and sometimes cable interruption has indicated earthquake action so far from land as not to have been felt by those on shore.   From want of knowledge of this kind of seismic effect, when three cables connecting Australia with Java were, in 1888, fractured simultaneously, the people of Australia called out the naval and military reserves, on the supposition that their sudden isolation indicated an operation of war. When it is remembered that this is by no means the only time a British colony has been suddenly cut off from communication with the rest of the world by the breaking of cables, the importance of being able to say whether this was brought about by natural or by artificial means cannot be over-estimated. Suitable instruments, wherever they are established, give information of great seismic disturbances, which may even take place at the antipodes of the place of observation. Hence they enable us to correct, confirm, and even to disprove the telegraphic information in our daily papers. From the enormous rate at which these earth messages pass through our earth we see that our ideas respecting the effective rigidity of our planet must be modified.

The records of the unfelt movements of the earth's crust often throw light upon unusual movements observed in the photograms from magnetographs, barographs, and other apparatus, and therefore instruments which yield these seismic records are indispensable adjuncts to all fully equipped observatories.

To what extent a more extended knowledge of the conditions under which diurnal waves are produced may prove beneficial is as yet problematical. If they are the result of evaporation and condensation we see in them phenomena bearing upon agriculture and forestry. Diurnal changes in inclination in exceptional cases may be so marked as to produce slight variations in the zero point of delicate balances and pendulums. Changes of this nature of sufficient magnitude to attract the attention of those engaged in the determination of standard weights, delicate assaying, or gravitational experiments are more likely attributable to pulsations, earth tremors, and sudden irregular displacements of the vertical. In reports upon the determination of standard weights, notes occur showing that there have been times when balances did not behave with regularity.

Whilst in Japan I placed an Oërtling and a Bunge assay balance in parallel east-west positions upon a massive stone column, and left them on the swing for a period of

U

twenty days. It was seldom that either of them was absolutely at rest, and the former, which was the lighter of the two, showed the most motion. This balance, which during a day usually crept through half a division on the ivory scale, had a period of forty-one seconds. Sometimes it was found performing complete swings in intervals of from seventeen to sixty seconds. Slower motions might take fifty minutes. These oscillations were not about the same zero, and the zero point might change within a few minutes. Very often the balances would start from rest in the same direction and at the same time. Movements occurred during tremor storms and when tremors were absent, when the barometer was high and when the weather was calm. Both balances were observed to be absolutely at rest during a heavy gale and when the barometer was low.

At present we have no means of saying that the erratic behaviour of the two balances here described was really the result of *earth* movements. The main fact is that such movements have at certain times a marked existence, and even if we were to determine a zero point before every weighing, errors of a serious magnitude might creep into the assayer's determinations. If future investigation shows them to be connected with ' earth ' tremors, then a knowledge of these will indicate the times at which weighing can be most satisfactorily performed.

These remarks on weighing have evidently a bearing upon gravitational experiments, in which the swinging of a pendulum may, through imperfectly understood causes be accelerated or retarded.

The addition or removal of a comparatively small load to or from one side of an observing station may often produce a marked change in level.

Mr. H. C. Russell, writing from Sydney, tells me that when Mr. E. F. J. Love was swinging the Kater pendulums at that place, it was thought that the old framework on which the pendulums swung was very unstable. This led Mr. Russell to fit up a design to test the point. The cellar in which the experiments were made had a concrete cement floor six inches thick. On one side of the pendulums there was a wall two feet six inches thick, and on the other the transit circle pier six feet thick and sixteen feet long. The result of the experiments showed that the weight of a man produced large deflections, whilst the effect of a 7 lb. weight was quite appreciable. At Oxford, Professor H. H. Turner and the author measured the deflections produced in a similar pier by a man and two boys moving on a concrete floor to different points round its base. A horizontal pendulum may be set swinging for the purpose of determining its period by standing near the base of its supporting pier, whilst the same may be adjusted by shifting the position of a 10 lb. weight on a concrete floor carrying the pier. Artificial shifting of loads near to the base of instruments may be avoided, but it is more difficult to avoid those changes in load which may take place unequally on two sides of an instrument and which are the result of natural actions. One means of at least mitigating these is to locate an observing station so that effects accompanying solar radiation shall be as pronounced upon one side of it as upon another. If a rule of this description is violated —as, for example, by placing an observatory so that on its eastern side it has grass land, on the other a bare surface —the difference in the loads removed daily from the two sides of such an observatory by evaporation might amount to and even exceed 12 lb. per square yard, a quantity sufficiently large to produce considerable changes in level.

   Another interesting remark of Mr. Russell is that in determining differences in longitude in New South Wales by transit circle observations, nearly always on one or two days out of six or seven, when simultaneous observations were made, these would contradict each other, although the observers vouched for the satisfactory quality of work on every night. As a probable explanation of this an erratic change in the position of the vertical is suggested.

   At the Edinburgh Royal Observatory a bifilar pendu-

lum has been established, not for the purpose of recording
earthquakes, but to throw light upon those changes in the
nadir reading—which is the index point in connection with
the declination of stars—that cannot be accounted for by
temperature, change in the instrument, or errors of observation.

Astronomical, spectroscopic, and photographic work
and observations dependent upon reflection from a basin of
mercury have repeatedly been interfered with, not only by
artificially produced vibrations, but by tremors having a
natural origin. In 1887, when Professor Todd came to
Japan to observe an eclipse of the sun, the chief portion
of the work was to obtain photographs of the corona. By
means of a heliostat and a forty foot lens the image of
the sun was thrown upon the photographic plate. The
apparatus was installed upon solid stone columns, but we
learn from Professor W. K. Burton that at times it seemed
impossible to obtain a steady image, a possible explanation
for which is that at the time of these observations earth
tremors were pronounced.

Earth Tremors and Fire Damp

When we remember that every year the deaths which
occur at collieries in England amount to about 1,000,
twenty-five per cent. of which are due to explosions of
fire damp, and that these are accompanied by injuries to
many others and an enormous loss of property, any investigation which is likely to minimise these disasters cannot fail in being welcome. That there is a relationship
between the escape of fire damp and certain meteorological conditions has been so far recognised that the
Mines Regulation Act demands that a barometer and a
thermometer be placed near to the entrance of a mine,
and that the changes observed in these instruments should
be accompanied by certain precautionary measures. The
character of the facts which have led Governments to
legislate on these matters may be judged of from the

following extracts taken from the writings of M. Chesneau and the report of the Austrian Fire Damp Commission ('Annales des Mines,' 1888, vol. ix. 1886, p. 258; vol. xiii. p. 389).

M. Le Chatelier, after a critical examination of the investigations made by Galloway between 1868 and 1873, arrives at the conclusion that it is doubtful if variations in atmospheric pressure have any relationship with the escape of gas.

Mr. Schöndorf, who made observations in the Saar Basin, concluded that barometric fluctuations directly affected the escape of gas from the goaf. M. Nasu, by carefully examining the gas issuing from a particular bed, found that it increased with a barometric fall; but, as pointed out by M. Chesneau, it is likely the increase may have been solely due to a greater escape from the area enclosed by stoppings rather than an increased rate of distillation from the coal.

The experiments of M. Hilt led to the conclusion that gas increased with a barometric fall, and *vice versâ*, but the examination of M. Hilt's results by Messrs. Mallard and Le Chatelier showed that great barometric falls only correspond with the appearance of small quantities of gas, whilst in regions cut off from goaf the correspondence was barely evident.

In the case of one considerable fall, the gas decreased at one mine, while at another it increased.

The conclusions arrived at by Mr. Köhler at certain mines in Silesia were as follows :

1. The quantity of gas diminishes with a rise of the barometer, and *vice versâ*.

2. The quantity increases proportionately to the rate at which the barometer falls, and *vice versâ*.

3. The quantity of gas disengaged is not absolutely dependent on the height of the barometer.

4. If the barometer rises rapidly and after that very gently, or remains steady at its maximum, a small increase of gas takes place; inversely, if

it falls rapidly and then gently rises or remains long at a minimum, a diminution in the quantity of gas commences.

The quantity of gas was determined by chemical analysis.

The closest agreements between barometric fluctuations and the disengagement of gas occur at the mines where the old workings cover an extensive area. Mr. Köhler also made experiments by hermetically sealing the downcast and producing a depression by the revolutions of a fan, with the result that the quantity of gas was considerably increased, even when there was no communication with the goaf.

From the researches of the Austrian Fire Damp Commission, in five districts not containing old workings practically no connection was observed between the liberation of fire damp and fluctuations in barometric pressure. It was, moreover, shown by experiment that the gas was contained in the coal under considerable pressure (in one case as high as 9·2 atmospheres), from which it might be inferred that slight barometric changes would produce no sensible effect upon the escape of gas. It was also observed that the volumes of gas collected from boreholes did not vary with atmospheric pressure.

The gas coming from old workings closely followed the barometric curve.[1]

The conclusion is that a local barometrical fall directly affects the escape of gas from old workings and goaves, whilst it has in the majority of instances but little effect upon the escape of gas from coal. From Mr. Köhler's second observation we may infer that it is possibly connected with the steepness of the barometric gradient.

Next we will turn to the tromometric records which have been obtained at collieries.

In 1883 I established a number of instruments in the Takashima Colliery near Nagasaki in South Japan, the workings of which are partly beneath the bed of the Pacific

---

[1] *Trans. Fed. Inst.*, vol. iii. p. 534.

Ocean. The movements looked for were tremors, disturbances due to the bending of the superincumbent strata by the rising and falling of the tide, and earthquakes. Only a few observations were made when a ' fall ' occurred, and the instruments have remained buried ever since.[1] Reference is made to these investigations by Mr. M. Walton Brown in a paper ' On the Observation of Earth Shakes or Tremors, in order to foretell the issue of sudden outbursts of fire damp ' ; [2] also by M. G. Chesneau.

Mr. Walton Brown points out the fact that the frequency of earthquakes in Great Britain and the fatal explosions of gas in collieries have each been greater during the winter months. Although a seismograph was esta-blished at Marsden and a committee appointed to inquire into the observation of earth tremors in mining districts, nothing has been done in England, so far as I am aware, to determine whether there is any relationship between the movements considered in this chapter and the escape of fire damp.[3]

In France, however, at Douai, the liberation of fire damp in connection with the movements of tromometers has received careful attention, an account of which is given by M. Chesneau. By means of a Pieler's lamp the gas in the returns was measured daily at 6 A.M., at which time, on account of work ceasing at 5.30 A.M., the volume of gas was as far as possible independent of the quantity of coal being extracted.

Barometric observations were made on the surface and underground, whilst tromometric observations were made with a *tromomètre normal* consisting of a pendulum 1·50 metres long the style of which was observed with a microscope. My analysis of the results given by M. Chesneau seems to show that for the particular collieries where the

[1] ' On Earth Pulsations and Mine Gas,' by J. Milne, *Trans. Fed. Inst. Min. Eng.*, June 2, 1893, and *Report to the British Association on the Volcanic Phenomena of Japan*, 1886, p. 413.

[2] *Trans. N. E. Inst. Min. Eng.*, vol. xxxiii. p. 179.

[3] *Ibid.*, ' Seismometer used at Marsden,' vol. xxxvii. pl. viii. et vol. xxxvii. p. 55.

percentage of gas in the returns was measured from June to the end of October the gas seldom reached 1 per cent., whilst in November, December, February, and March it was usually above 1 per cent. For the other months, January, April, and May, no returns were given. The data, such as exist, are sufficient to show that at this particular mine the escape of gas followed the winter rule, which indicates that there may be a general coincidence in the times at which tremors are most frequent and the development of mine gas most pronounced. A still closer relationship, however, exists between these phenomena, and the discussion of M. Chesneau's results shows microseismic movements to be more clearly related to the escape of gas than to barometric movements. On some occasions this relationship between the three phenomena has been extremely well marked, as, for example, on December 8, 1886.

A point in connection with this which, although not referred to by M. Chesneau, can hardly have escaped his attention, is that although the increase in microseismic movements, the increase in gas, and the barometric fall commenced simultaneously, the microseismic movements reach a maximum about six hours before the gas reaches a maximum, whilst the lowest point of the barometric curve occurs even twelve hours later, or eighteen hours after the maximum of the tromometric movements.

To know whether these earth movements or even their maxima are always somewhat in advance of the escape of fire damp is a matter for future experiment and, in my opinion, can be determined only by the use of instruments yielding a continuous automatic record. The only other work with which I am acquainted, and which bears on the matter now under consideration, is a comparison between the monthly curves of microseismic activity in Italy and a number of explosions of fire damp recorded in Germany. These are arranged as a monthly curve, and show a close relationship with the microseismic curves, which in the times of their maxima and minima show a close agreement throughout the Italian peninsula.

Artificially Produced Vibrations

Astronomers and physicists throughout the world are well aware of the hours that have been lost, the errors that have been occasioned, and the annoyance that has been created in consequence of the existence of elastic tremors created by trains, carriages, and traffic. Sir George Airy, to escape the effects of vibrations on Bank Holidays when Greenwich Park was filled with merry-makers, suspended the vessel containing the reflecting surface of mercury by indiarubber bands, a method now followed by surveyors and photographers who work in cities where vibrations are pronounced. General H. S. Palmer, when observing the transit of Venus in New Zealand, protected himself against the disturbing influence of passing trains 400 yards distant by digging entrench-ments around the piers of his instruments about four feet in depth.

Professor H. M. Paul, in an account of experiments made in Washington in connection with the examination of proposed sites for the U. S. Naval Observatory, tells us that at a distance of about a mile from a railroad the effect of trains upon the reflecting surface of mercury was visible for one minute. At another station, the intensity of the vibrations seemed to be reduced though not entirely cut off by a ravine fifty or sixty feet deep. Carriages pro-duced serious disturbances at distances of 300 feet. The ground through which these movements were transmitted was clay and gravel.

My own experiments show that a ball 1,710 lb. in weight falling about thirty-five feet produced vibrations visible to the eye on a surface of mercury at a distance of from 400 to 600 feet. A somewhat heavier ball falling forty feet upon a clay-like tuff rock did not produce an effect recordable by a seismograph at a distance greater than twenty feet. With a Perry tromometer the effect of trains through soil lying above chalk can be seen at a distance of about one mile. The firing of a heavy gun on ship-board at a distance of six miles may cause a spot of light three feet distant from the instrument to be deflected through a distance of one foot. But this deflection takes place when the sound is heard, and is probably due to the mechanical vibration produced by the air-wave upon the building shaking the wall to which the instrument is attached.

Vibrations and Jolts on Trains

In a train we experience not only elastic vibrations in great measure due to the yielding óf the carriage on its springs, but in consequence of the wearing of tyres and bearings, variation in gauge, collisions at facing points, the changes in character of ballast and sleepers, and from other irregularities, we also experience forced displacements or jolts. All of these—which vary in different parts of a carriage or a train, and which are largely in evidence at high speeds and in rear carriages—can be recorded by means of an instrument developed from a seismograph. Under all circumstances irregularities of motion are indicated as excrescences in the general diagram, which therefore not only gives automatically a record of the condition of a line, but of the speed of the train and the duration of the stoppages. Each jolt that is recorded indicates loss of energy, which may not only annoy a traveller, but be propagated to a distance of a mile on each side of a line and disturb delicate instruments.

A diagram taken on a locomotive indicating a pronounced fore and aft motion tells us that the balancing of the wheels is such that energy is being lost in what is equivalent to the stopping and starting of a load. By correcting this form of motion it has been found in Japan that there is a marked decrease in the consumption of fuel, the safe speed of travelling can be increased, whilst workshop repairs are reduced.

Vibrations of Bridges, Buildings, and Steamships

With bridges, buildings, and steamboats we have to deal with vibrations of an elastic character.

In ordinary practice the test to which iron girders are usually subjected is the maximum vertical deflection which they show under moving and stationary loads. The record from a seismograph shows not only this but also the upward deflections of a bridge due to resilience, the elastic vibratory motions which outrace an approaching train, the compound longitudinal bendings, the transverse displacements, and the natural periodic swing which continues long after a train has passed.

The character of these diagrams varies with that of the girders, the state of the track, the lateral freedom of the travelling load, its weight, and the rate at which it passes. The most pronounced movement is in a transverse direction, and on account of the slowness of the natural period in this direction, the cross swing of a bridge may be accentuated by the synchronisms of the impacts of the travelling load. The greatest swing experienced is therefore not necessarily produced by the heaviest train, but by a particular carriage travelling at a particular speed. In the case of long girders a similar condition may accentuate the vertical motion. Sometimes it is noticed that this bending is greatest, not when the locomotive passes a bridge, but by comparatively light carriages near the end of a train. A vibration metre established on a bridge would not only record deflections due to wind and traffic, but it would record the time at which these disturbances took place and the speed at which trains had passed ('Engineering,' January 24, 1896, p. 111).

The origin of the vibrations which occur in factories and other buildings and those in high speed steamships may usually be traced to a want of balance between the rotating and reciprocating portions of steam or other motors. The vibrations in buildings and factories, although often annoying, are characterised by their rapidity and their smallness, their amplitudes being measured by quantities usually lying between ·01 and ·001 of an inch.

One curious feature often to be observed is their pulsatory character. They are also unequally distributed in a structure. In the case of a tall building they may be marked at about two-thirds its height.

In torpedo destroyers the elastic vibrations are very pronounced at particular speeds. They may have a range exceeding one inch, and occur with such rapidity that the acceleration may exceed that due to gravity, and cause objects to dance upon a table (' Engineering,' March 13, p. 337).

SEISMOLOGICAL LITERATURE

A GENERAL outline of the character of seismological literature may be found in the volume on ' Earthquakes ' published in the International Scientific Series. In 1895, in consequence of a fire, the writer lost, amongst other things, his library, which included some 1,500 books and papers relating to seismology. The result of this has rendered references to work done outside Japan occasionally incomplete. The following list of papers published in Japan, the contents of many of which are epitomised in reports to the British Association between the years 1881 and 1898, will give some idea of the nature of the investigations carried out in that country. A general outline of the work carried out in Italy will be found in a British Association report for 1898.

TRANSACTIONS OF THE SEISMOLOGICAL SOCIETY OF JAPAN

GENERAL INDEX, VOLS. 1-20

The letters S. J. refer to the Seismological Journal of Japan, a publication which may be regarded as a continuation of the Transactions.

APPENDIX

REPRODUCED, WITH ADDITIONS, FROM 'EARTHQUAKES,'
INT. SCI. SERIES.

—◆◇◆—

LIST OF THE PRINCIPAL BOOKS, PAPERS, PERIODICALS, WHICH ARE
REFERRED TO IN THE PRECEDING PAGES.

*For a more complete bibliography of earthquakes refer to Mallet's
catalogue of works given in his report to the British Association in
1858.*

A True and Particular Relation of the Dreadful Earthquake which
happened at Lima, &c. (1746). 1768.
Abbadie, M. A. d'. *See* p. 248.
Abbot, Gen. H. L. On the Velocity of Transmission of Earth Waves
Am. Jour. Sci. XV., March 1878.
— Shock of the Explosion at Hallet's Point, Nov. 14, 1876. *Battalion
Press.*
Agamennone, Dott. G. *See* p. 126.
Alexander, Prof. T. See *Trans. Seis. Soc. of Japan.*
American Journal of Science.
Annali del reale osservatorio meteorologico Vesuviano.
Annual Register, The.
Anonymous, A Chronological and Historical Account of the most
Memorable Earthquakes in the World, &c. 1750.
— A Vindication of the Bishop of London's Letter occasioned by the
Late Earthquake. 1750.
— Phenomena of the Great Earthquake of Nov. 1, 1755.
— Serious Thoughts occasioned by the Late Earthquake at Lisbon.
1755.
Asiatic Society of Japan, Transactions of.
Ayrton, Prof. W. E. *See* Perry, J.

Bárceno, M. Estudio del Terremoto (May 17, 1879) Mexico. 1879.
Beiträge zur Geophysik. 3 Vols. Dr. G. Gerland.
Beke, Dr. C. T. Mount Sinai a Volcano.

Bissett, Rev. J.   A Sermon (on account of the Earthquake at Lisbon,
    Nov. 1, 1755).   1757.
Bittner, A.   Beiträge zur Kenntniss des Erdbebens von Belluno vom 29.
    Juni 1873.
— Sitzungsb. der K. Akad. d. Wissensch., lxix. II. Abth., 1874.
Bollettino della Società Sismologica Italiana.   Vols. I.–III.
Bollettino del Vulcanismo Italiano.
Boué, Dr. A.   Ueber das Erdbeben welches Mittel-Albanien im October
    d. J. so schrecklich getroffen hat.   Die K. Akad. d. Wissenschaften,
    Nov. 1851.
— Parallele der Erdbeben, des Nordlichtes und des Erdmagnetismus.
— Ueber die Nothwendigkeit die Erdbeben und vulcanischen Er-
    scheinungen genauer als bis jetzt beobachten zu lassen.   Die K.
    Akad. d. Wissenschaften, 1851 and 1857.
Bouguer, M.   Of the Volcanoes and Earthquakes in Peru.
British Association, Reports of.
Brunton, R. H.   Constructive Art in Japan.   Trans. Asiatic Soc. of
    Japan, II. and III., Pt. 2.
Bryce, J.   Report to British Association, 1841.
Buffour, M.   The Natural History of Earthquakes and Volcanoes.

C. H.   A Physical Discussion of Earthquakes, &c.   1693.
Cancani, Dott. A.   See p. 125.
Canterbury, Thomas, Lord Archbishop of.   The Theory and History of
    Earthquakes.
Casariego, E. A.   See Trans. Seis. Soc. of Japan.
Cawley, G.   Some Remarks on Construction in Brick and Wood, &c.
    Trans. Asiatic Soc. of Japan, VI. Plate ii.
Chaplin, Prof. W. S.   An Examination of the Earthquakes recorded at
    the Meteorological Observatory, Tokio.   Trans. Asiatic Soc. of
    Japan, VI. Part ii.
Comptes Rendus.
Credner, H.   Das Dippoldiswalder Erdbeben vom Oktober 1877.
— Zeitschr. f. d. Naturwiss. f. Sachsen u. Thüringen.
— Das Vogtländisch-erzgebirgische Erdbeben, 23. Nov. 1875.
— Zeitsch. f. d. gesammt. Naturwissenschaften, xlviii., Oktober.

Dan, T.   See Trans. Seis. Soc. of Japan.
Darwin, Charles.   Researches on Geology and Natural History.
— Geological Observations.
Darwin, G. H.   Reports on Lunar Disturbance of Gravity to British
    Association, 1881.   1882.
— The Tides.   1898.
Davison, Dr. Charles.   On the Theory of Vorticose Earthquake Shocks.
    Geolog. Mag., Vol. IX., 1882, pp. 257–265.
— On a Possible Cause of the Disturbance of Magnetic Compass Needles
    during Earthquakes.   Geolog. Mag., Vol. II., 1885, pp. 210–211.
— On the Existence of Undisturbed Spots in Earthquake-shaken Areas.
    Birm. Phil. Soc. Proc., Vol. V., 1886, pp. 57–60.

Davison, Dr. Charles. Note on M. Ph. Plantamour's Observations by
means of Levels on the Periodic Movements of the Ground at
Sècheron, near Geneva. *Phil. Mag.*, Feb. 1889, pp. 189-199.
— On the Study of Earthquakes in Great Britain. *Nature*, Vol. XLII.,
1890, pp. 346-349.
— On the British Earthquakes of 1889. *Geolog. Mag.*, Vol. VIII., 1891,
pp. 57-67, 306-316, 364-372.
— On the British Earthquakes of 1890, &c. *Geolog. Mag.*, Vol. VIII.,
1891, pp. 450-455.
— On the Inverness Earthquakes of Nov. 15 to Dec. 14, 1890. *Quart.
Jour. Geolog. Soc.*, Vol. XLVII., 1891, pp. 618-632.
— Record of Observations on the Inverness Earthquake of Nov. 15, 1890.
Birm. Phil. Soc. Proc., Vol. VIII., 1892.
— On the Nature and Origin of Earthquake Sounds. *Geolog. Mag.*,
Vol. IX., 1892, pp. 208-218.
— On the British Earthquakes of 1891. *Geolog. Mag.*, Vol. IX., 1892,
pp. 299-305.
— On the Annual and Semi-annual Seismic Periods. *Phil. Trans.*,
1893 A, pp. 1107-1169.
— On the British Earthquakes of 1892. *Geolog. Mag.*, Vol. X., 1893,
pp. 291-302.
— Note on the Quetta Earthquake of Dec. 20, 1892. *Geolog. Mag.*,
Vol. X., 1893, pp. 356-360.
— Report of Brit. Assoc. Com. on Earth Tremors, 1893. *Brit. Assoc.
Report*, 1893, pp. 287-309. (Bifilar Pendulum designed by Mr. H.
Darwin, pp. 291-303.)
— Report of Brit. Assoc. Com. on Earth Tremors, 1894. *Brit. Assoc.
Report*, 1894, pp. 145-154. (The Greek Earthquake Pulsations of
April, 1894, pp. 146-154.)
— On the Leicester Earthquake of Aug. 4, 1893. *Royal Soc. Proc.*,
Vol. LVII., 1895, pp. 87-95.
— Bifilar Pendulum for Measuring Earth-tilts. *Nature*, Vol. L., 1894,
pp. 246-249.
— On the Velocity of the Constantinople Earthquake Pulsations of
July 10, 1894. *Nature*, Vol. L., 1894, pp. 450-451.
— The Horizontal Pendulum. *Natural Science*, Vol. VIII., 1896, pp.
233-238.
— On the Comrie Earthquake of July 12, 1895, &c. *Geolog. Mag.*,
Vol. III., 1896, pp. 75-78.
— On the Diurnal Periodicity of Earthquakes. *Phil. Mag.*, Dec. 1896,
pp. 463-476.
— On the Exmoor Earthquake of Jan. 23, 1894. *Geolog. Mag.*, Vol. III.,
1896, pp. 553-556.
— Note on an Error in the Method of Determining the Mean Depth of
the Ocean from the Velocity of Seismic Sea-waves. *Phil. Mag.*,
Jan. 1897, pp. 33-36.
— On the Effect of the Great Japanese Earthquake of 1891 on the
Seismic Activity of the adjoining Districts. *Geolog. Mag.*, Vol. IV.,
1897, pp. 23-27.

310 SEISMOLOGY

Davison, Dr. Charles. On the Distribution in Space of the Accessory
Shocks of the Great Japanese Earthquake of 1891. *Quart. Jour.
Geolog. Soc.*, Vol. LIII., 1897, pp. 1–15.
— On the Pembroke Earthquakes of Aug. 1892, and Nov. 1893. *Quart.
Jour. Geolog. Soc.*, Vol. LIII., 1897.
— On the Distribution of Earthquakes in Japan during the Years 1885–
1892. *Geograph. Jour.*, Nov. 1897.
Diffenbach, F. Plutonismus und Vulkanismus in der Periode von 1868–
1872, und ihre Beziehungen zu den Erdbeben im Rheingebiet.
Doelter, C. von. Ueber die Eruptivgebilde von Fleims, nebst einigen
Bemerkungen über den Bau älterer Vulcane.
— lxxiv. Band. d. Sitzungsb. d. K. Akad. d. Wissensch., I. Abth., Dec.
Heft, Jahrg. 1876.
Doolittle, Rev. T. Earthquakes Explained and Practically Improved,
&c. 1693.
Doyle, P. See *Trans. Seis. Soc. of Japan.*
Dutton, Capt. C. E. See p. 126.

Ehlert, Dr. R. See p. 248.
Emerson, Prof., B.A. Review of Von Seebachs' Earthquake of March 6,
1872. *Am. Jour. Sci.*, Series III.
Ewing, Prof. J. A. Earthquake Measurement. A memoir published by
the Tokio University. 1883.
— See *Trans. Seis. Soc. of Japan.*

Falb, R. Gedanken und Studien über den Vulcanismus, &c. 1875.
— Grundzüge zu einer Theorie der Erdbeben und Vulkanausbrüche.
— Das Erdbeben von Belluno. ' Sirius,' Bd. VI., Heft ii.
Flamstead, J. A Letter concerning Earthquakes. 1693.
Forel, F. A. Les Tremblements de Terre (Suisse). *Arch. des Sciences
Physiques et Naturelles*, VI. p. 461.
— Tremblement de Terre du 30 Décembre 1879.
Fouqué, F. See p. 126.
Fuchs, Karl. Vulkane und Erdbeben.
— *Die Vulkanischen Erscheinungen der Erde.*

Garcia, J. C. See *Trans. Seis. Soc. of Japan.*
Geinitz, Dr. E. Das Erdbeben von Iquique am 9. Mai 1877, &c. *Die
K. Leop.-Carol.-Deutschen Akademie der Naturforscher*, Band xl.,
Nr. 9.
Gentleman's Magazine, The.
Geographical Society, Proceedings of.
Geological Society, Proceedings of.
Girard, Dr. H. Ueber Erdbeben und Vulkane. 1845.
Gray, T. See *Trans. Seis. Soc. of Japan.*
— On Instruments for Measuring and Recording Earthquake Motions.
Phil. Mag., Sept. 1881.
— On Recent Earthquake Investigation. *The Chrysanthemum*, 1881.
Guiscardi, Prof. G. Notizie del Vesuvio. 1857.

Guiscardi, Prof. G. Il terremoto di Casamicciola del 4 Marzo. 1881.
Hales, S,, D.D., F.R.S. Some Considerations on the Causes of Earthquakes. 1750.
Hamilton, Sir W. Observations on Mount Vesuvius, Mount Etna, &c. 1774.
Hattori, I. Destructive Earthquakes in Japan. *Trans. Asiatic Soc. of Japan*, V. Plate i.
Heim, Prof. A. Les Tremblements de Terre et leur Etude Scientifique. 1880.
— Prof. A. Die Schweizerischen Erdbeben in 1881-1882.
Hoeffer, Prof. H. Die Erdbeben Kärntens und deren Stosslinien. *Die Kais. Akademie d. Wissenschaften*, Band xlii.
Höfer, Prof. H. Das Erdbeben von Belluno, am 29. Juni 1873. *Sitzungsb. der K. Akad. d. Wissensch.*, I. Abth., Band lxxiv.
Hoff, K. E. A. von. Geschichte der durch Ueberlieferung nachgewiesenen natürlichen Veränderungen der Erdoberfläche. 1822.
Hooke, R., M.D., F.R.S. Discourses concerning Earthquakes.
Hopkins, William. Report to the British Association on the Geological Theories of Elevation and Earthquakes. 1847.
Horton, Rev. Mr. An account of the Earthquake which happened at Leghorn in Italy (Jan. 1742). 1750.
Humboldt, Alexander von. Cosmos.
— Travels.

Jeitteles, L. A. Bericht über das Erdbeben am 15. Januar 1858.
— Sitzungsberichte der mathem.-naturw. Classe d. K. Akad. d. Wissensch., xxxv. S. 511.
Judd, J. W., Prof. Volcanoes, What they Are, and What they Teach.

Knipping, E. Verzeichniss von Erdbeben wahrgenommen in Tokio, &c. *Mitt. d. Deutsch. Gesellsch. für Natur- und Völkerkunde Ostasiens*, Heft 14.
— See *Trans. Seis. Soc. of Japan*.
Knott, C. G. See *Trans. Seis. Soc. of Japan*.
— On Lunar Periodicities in Earthquake Frequency. *Proc. Royal Soc.*, London, Vol. LX., 1897.

Lasaulx, A, von. Das Erdbeben von Herzogenrath am 22. October 1873.
Lemery, M. A Physico-Chemical Explanation of Subterranean Fires, Earthquakes, &c.
Lescasse, M. J. Etude sur les Constructions Japonaises, &c. *Mémoires de la Société des Ingénieurs Civils*.
Lévy, M. *See* p. 126.
Lister, M., M.D., F.R.S. Of the Nature of Earthquakes.
Little, Rev. J. Conjectures on the Physical Causes of Earthquakes and Volcanoes. 1820.

Mallet, R. The Neapolitan Earthquake, Vol. II. *Reports to the British Association*, 1850, 1851, 1852, 1854, 1858, 1861.

312 SEISMOLOGY

Mallet, R. Secondary Effects of the Earthquake of Cachar. *Proc. Geolog. Soc.*, 1872.
— Dynamics of Earthquakes. *Trans. Royal Irish Acad.* 1846.
Michell, Rev. J. Conjectures Concerning the Cause and Observations upon the Phenomena of Earthquakes. 1760.
Milne, David. Reports to British Association, 1841, 1843, 1844.
Milne, John. See *Trans. Seis. Soc. of Japan:*
— On Seismic Experiments (with T. Gray, B.Sc., F.R.S.E.). *Trans. Royal Soc.* 1882.
— On Seismic Experiments (with T. Gray, B.Sc., F.R.S.E.). *Proc. Royal Soc.* No. 217, 1881.
— Earthquake Observations and Experiments in Japan (with T. Gray, B.Sc., F.R.S.E.). *Phil. Mag.*, Nov. 1881.
— On the Elasticity and Strength Constants of certain Rocks (with T. Gray, B.Sc., F.R.S.E.). *Jour. Geolog. Soc.*, 1882.
— A Visit to the Volcano of Oshima. *Geolog. Mag.*, Dec. 2, Vol. IV., pp. 193–197, 255.
— On the Form of Volcanoes. *Geolog. Mag.*, Dec. 2, Vol. V., and Dec. 2, Vol. VI.
— Note upon the Cooling of the Earth, &c. *Geolog. Mag.*, Dec. 2, Vol. VII., p. 99.
— Investigation of the Earthquake Phenomena of Japan. *Eighteen Reports Brit. Assoc.*, 1881 to 1898.
— A Large Crater. *Popular Science Review.*
— The Volcanoes of Japan (a series of articles). *Japan Gazette.*
— Earthquake Literature of Japan (a series of articles). *Japan Gazette.*
— The Earthquake of Dec. 23, 1880. *The Chrysanthemum*, 1881.
— Earthquake Motion. *The Chrysanthemum*, 1882.
— Seismology in Japan. *Nature*, Oct. 1882.
— Earth Movements. *The Times*, Oct. 12, 1882.
— Cruise in the Kurile Islands. *Geolog. Mag.*, Dec. 2, Vol. IV., p. 337.
— Geographical Distribution of Volcanoes. *Geolog. Mag.*, Dec. 2, Vol. VII., p. 166.
— Construction in Earthquake Countries. *Proc. Inst.*, *C.E.*, Vol. LXXXIII., 1885–6.
— Building in Earthquake Countries. *Proc. Inst. C.E.*, Vol. C., 1889–90.
— Movements of the Earth's Crust. *Geograph. Jour.*, March 1896.
— Sub-Oceanic Changes. *Geograph. Jour.*, Aug. and Sept. 1897.
— Recent Advances in Seismology. *Royal Inst. Lecture*, Feb. 1897.
— The Great Earthquake of 1891. Published with W. K. Burton in Japan.
— Causes of Earthquakes. *Science Society*, Tokio.
— Earth Pulsations and Mine Gas. *Inst. Mining Eng.*, June 1893.
Mohr, Dr. F. Geschichte der Erde. 1875.

Naturkundig Tijdschrift voor Nederlandsch Indie. 1875–1880.
Naumann, Dr. E. Ueber Erdbeben und Vulkanausbrüche in Japan. *Mitt. d. Deutsch. Gesellsch. für Natur- und Völkerkunde Ostasiens*, Heft 15.
Newcomb, Prof. S. *See* p. 126.

Noggerath, Dr. J. Die Erdbeben vom 29. Juli 1846 im Rheingebiet, &c.
— Die Erdbeben im Vispthale (1855).
— Die Erdbeben im Rheingebiet in den Jahren 1868, 1869, 1870.
— Jahrgänge d. Verhandlungen d. Natur. Vereins für 1870. *Rheinland u. Westphalen*, xxvii.

Oldham, Dr. Secondary Effects of the Earthquake of Cachar. *Proc. Geolog. Soc.* 1872.
— Thermal Springs of India. *Memoirs of Geolog. Survey of India*, XIX. Plate 2.
— A Catalogue of Indian Earthquakes. *Memoirs of Geolog. Survey of India*, XIX. Plate 3.
— The Cachar Earthquake. *Memoirs of Geolog. Survey of India*, XIX. Plate 1.

Palmer, Col. H. S. See *Trans. Seis. Soc. of Japan.*
Palmieri, Prof. L., e Scacchi, A. Della Regione Volcanica del Monte Vulture, e del Tremuoto ivi avvenuto nel dì 14 Agosto 1851. 1852.
— Annali del reale Osservatorio Meteorologico Vesuviano.
— Il Vesuvio, il Terremoto d' Isernia e l'eruzione sottomarina di Santorino. *R. Accad. d. Sci. Fis. e Mat. di Napoli*, iv. 1866.
— Sul recente Terremoto di Corleone. *R. Accad. d Sci. Fis. e Mat.*, v. 1876.
— Il Terremoto di Scio del dì 4 Aprile. *R. Accad d. Sci. Fis. e Mat. di Napoli*, v. 1881.
— Sul Terremoto di Casamicciola del 4 Marzo 1881. *R. Accad. d. Sci. Fis. e Mat. di Napoli.* 1881.
Paul, Prof. H. M. See *Trans. Seis. Soc. of Japan.*
Perrey, Prof. A. Earthquake Catalogue and Memoirs. (For list see Mallet's Report to British Association. 1858.)
— See *Trans. Seis. Soc. of Japan.*
Perry, J., and W. E. Ayrton. On a Neglected Principle that may be Employed in Earthquake Measurement.
— See *Trans. Seis. Soc. of Japan.*
Philosophical Magazine.
Pickering, Rev. R. An Address to those who have either retired or intend to leave Town under the Imaginary Apprehension of the Approaching Shock of another Earthquake. 1750.

Ray, J., F.R.S. A Summary of the Causes of the Alterations which have happened to the Face of the Earth.
Rebeur-Paschwitz, Dr. E. von. *See* p. 248.
Rockstroh, E. Informe de la Comision Científica del Instituto Nacional de Guatemala, nombrada por el Sr. Ministro de Instruccion Pública para el Estudio de los Fenómenos Volcánicos en el Lago de Tlopango. 1880.
Rockwood, Prof. C. G. Notes on Earthquakes. Annually in the *Am. Jour. Sci.*
— Japanese Seismology. *Am. Jour. Sci.*, XXII. Dec. 1881.
Romaine, W. A Discourse occasioned by the Late Earthquake. 1755.

Rossi, Prof. M. S. di.   Intorno all' odierna fase dei Terremoti in Italia, e segnatamente sul Terremoto in Casamicciola del 4 Marzo 1881. *Società Geografica Italiana.* 1881.
— La Meteorologia Endogena, 2 vols.
Royal Society, Transactions of.
Russell, H. C.   *See* p. 248.

Scacchi, A.   *See* Palmieri.
Schmidt, D. A.   *See* p. 125.
Schmidt, Dr. J. F.   Untersuchungen über das Erdbeben am 15. Januar 1858.
— Studien über Erdbeben. 1879.
— Die Eruption des Vesuv (1855). 1856.
Scrope, G. P.   Volcanoes.
Seebach.   Das mittle Deutsche Erdbeben (1872). *Mitt. der K.K. geograph. Gesellsch.*, II. Jahrg., 2. Heft, 1873.
Serpieri, Prof. A. C. S.   Nuove Osservazioni sul Terremoto avvenuto in Italia il 12 Marzo 1873. *Istituto Lombardo.* 1873.
— Il Terremoto di Rimini della notte 17–18 Marzo 1875.
— Documenti nuove e Riflessioni sul Terremoto della notte 17–18 Marzo 1875. *Meteorologia Italiana*, iv. 1875.
— Determinazione delle fasi e delle leggi del grande Terremoto avvenuto in Italia nella notte 17-18 Marzo 1875. *Istituto Lombardo.* 1875.
— Dell' influenza del Lume Solare sui Terremoti. *Istituto Lombardo.* 1882.
Sherlock, T., D.D. (Lord Bishop of London).   A Letter on the occasion of the late Earthquakes. 1750.
Shower, Rev. J., D.D.   Practical Reflections on the Earthquakes that have happened in Europe and America, &c. 1750.
Stukeley, Rev. W., M.D., F.R.S.   The Philosophy of Earthquakes, Natural and Religious, &c.   Plates 1, 2, and 3. 1756.
Sturmius, J. C.   A Methodical Account of Earthquakes.
Suess, E.   Die Erdbeben Niederösterreiches. *Die Kais. Akademic der Wissenschaften*, Bd. xxxiii.
— Die Erdbeben des südlichen Italiens. *Die Kais. Akademie der Wissenschaften*, Bd. xxxiv.

Tacchini, P.   *See* p. 125.

Volger, Dr. G. H.   Untersuchungen über das Phänomen der Erdbeben. 1857.

Wagener, Dr. G.   Bemerkungen über Erdbebenmesser und Vorschläge zu einem neuen Instrumente dieser Art. *Mitt. d. Deutsch. Gesellsch. für Natur- und Völkerkunde Ostasiens*, Heft 15.
— See *Trans. Seis. Soc. of Japan.*
Winchilsea, the Earl of.   A True and Exact Relation of the late Prodigious Earthquake and Eruption of Mount Etna. 1669.
Woodward, J., M.D., F.R.S.   Earthquake caused by some Accidental Obstruction of a Continual Subterranean Heat.

INDEX

PRINTED BY
SPOTTISWOODE AND CO., NEW-STREET SQUARE
LONDON